AF352821

Electromagnetic Scattering and Its Applications: From Low Frequencies to Photonics

Electromagnetic Scattering and Its Applications: From Low Frequencies to Photonics

Editors

Gian Luigi Gragnani
Alessandro Fedeli

MDPI • Basel • Beijing • Wuhan • Barcelona • Belgrade • Manchester • Tokyo • Cluj • Tianjin

Editors
Gian Luigi Gragnani
University of Genoa
Italy

Alessandro Fedeli
University of Genoa
Italy

Editorial Office
MDPI
St. Alban-Anlage 66
4052 Basel, Switzerland

This is a reprint of articles from the Special Issue published online in the open access journal *Electronics* (ISSN 2079-9292) (available at: https://www.mdpi.com/journal/electronics/special_issues/electromagnetic).

For citation purposes, cite each article independently as indicated on the article page online and as indicated below:

LastName, A.A.; LastName, B.B.; LastName, C.C. Article Title. *Journal Name* **Year**, *Volume Number*, Page Range.

ISBN 978-3-0365-2207-4 (Hbk)
ISBN 978-3-0365-2208-1 (PDF)

Cover image courtesy of Alessandro Fedeli, Gian Luigi Gragnani.

Contents

About the Editors

Gian Luigi Gragnani (Professor) graduated in Electronic Engineering from the University of Genoa, Genoa, Italy, in 1985. In the same year, he joined the Applied Electromagnetics Group at the Department of Biophysical and Electronic Engineering, University of Genoa. Since 1989, he has been responsible for the Applied Electromagnetics Laboratory. He has also cooperated with the Inter-University Research Centre for Interactions Between Electromagnetic Fields and Biological Systems (ICEMB), of which he was deputy director. He is currently Professor of Electromagnetic Fields at the Department of Electrical, Electronics, Telecommunication Engineering and Naval Architecture (DITEN) of the University of Genoa, and he teaches courses on antennas and on electromagnetic propagation. His current research interests are on electromagnetic scattering (both direct and inverse), microwave imaging, wideband antennas, computational electromagnetics and open source software, and electromagnetic compatibility.

Alessandro Fedeli (Assistant Professor) received his B.Sc. and M.Sc. degrees in Electronic Engineering from the University of Genoa, Genoa, Italy, in 2011 and 2013, respectively, where he also was awarded his Ph.D. degree in Science and Technology for Electronic and Telecommunications Engineering in 2017. He is currently Assistant Professor of Electromagnetic Fields at the Department of Electrical, Electronics, Telecommunication Engineering and Naval Architecture (DITEN) of the University of Genoa, and he teaches courses on guided propagation, electromagnetic imaging, and remote sensing. His main research interests include computational methods for the solution of forward and inverse electromagnetic scattering problems, microwave imaging, and processing techniques applied to Ground Penetrating Radar (GPR) systems. He is Member of the IEEE, of the IEEE Antennas and Propagation Society, the Italian Society of Electromagnetism, the Inter-University Research Centre for Interactions Between Electromagnetic Fields and Biological Systems (ICEMB), the European Geosciences Union, and the Italian Georadar Association.

Editorial

Electromagnetic Scattering and Its Applications: From Low Frequencies to Photonics

Alessandro Fedeli *and Gian Luigi Gragnani *

Department of Electrical, Electronic, Telecommunications Engineering, and Naval Architecture,
University of Genoa, Via all'Opera Pia, 11A, 16145 Genoa, Italy
* Correspondence: alessandro.fedeli@unige.it (A.F.); gianluigi.gragnani@unige.it (G.L.G.)

Citation: Fedeli, A.; Gragnani, G.L. .
Electronics **2021**, *10*, 2352.
https://doi.org/10.3390/
electronics10192352

Received: 22 September 2021
Accepted: 24 September 2021
Published: 26 September 2021

Publisher's Note: MDPI stays neutral with regard to jurisdictional claims in published maps and institutional affiliations.

Few research topics are as broad and pervasive as electromagnetic scattering. Undeniably, electromagnetic scattering phenomena are behind many wireless and radio devices. From radar systems to mobile telecommunications, and from medical devices to innovative materials, the study of electromagnetic interactions and scattering is fundamental to develop applications that exploit electromagnetic waves.

Although the initial research in the field dates back centuries, many unresolved theoretical issues are still there. At the same time, novel applications of electromagnetic scattering are continuously emerging. A detailed description of the many areas in which scattering is involved is well beyond our scope. Based on the wavelength, applications may be very different, as well as the techniques used to analyze and simulate the electromagnetic propagation.

Starting from low-frequency problems and embracing the whole spectrum up to optics and photonics, the goal of the present Special Issue is to provide a comprehensive collection of state-of-the-art papers dedicated to electromagnetic scattering theory and applications.

Electromagnetic scattering from the so-called canonical structures is one of the subjects covered in this Special Issue. In particular, a theoretical framework is presented in [1] to provide a description of scattering phenomena involving dielectric and conducting spheres by decomposing the total field in the outer space in terms of inward and outward electromagnetic fields, rather than in terms of incident and scattered fields, as in the classical Lorenz–Mie formulation. The method can provide an intuitive physical interpretation of electromagnetic scattering in terms of impedance matching and resonances.

Of course, when the structures under consideration become more complex, the development of proper computational techniques is necessary to simulate and analyze the electromagnetic scattering effects. An example is the study of scattering problems in the presence of bianisotropic materials and metamaterials. In this regard, reference solutions obtained by means of finite-elements approximations are presented in [2]. In this paper, not only the well-posedness and the finite-element approximability in the three-dimensional and time-harmonic case are discussed, but also three example problems are considered: one, in free space, deals with scattering from plasmonic gratings that exhibit bianisotropy, while the other two deal with bianisotropic obstacles inside waveguides.

The propagation of electromagnetic waves in magnetized plasma is another research topic that deserves particular attention. In this Special Issue, this problem is addressed in [3], where a numerical method based on a modified precise-integration time-domain formulation is illustrated. It is shown that the method can provide an evident reduction in the execution time by using a larger simulated time step; meanwhile, the computational error of the presented algorithm is also lower than those of the formulations based on the FDTD scheme, especially in the high-frequency range.

In some applications it is important to study the scattering effects (e.g., the radar cross section) of large structures with details of different scales. In this situation, the adoption of full-wave numerical methods for the whole targets under test may require very high computational resources. A possible solution is represented by the use of hybrid techniques,

1

in which full-wave methods may be combined with approaches based on high-frequency approximations. This is the topic of the work presented in [4], where the finite volume time domain technique is hybridized with an asymptotic solution strategy based on physical optics and the equivalent current method.

Electromagnetic band gap structures (EBGs) [5], despite having been introduced more than twenty years ago [6], still continue to have a significant impact in microwave and antenna engineering research. This topic is considered in [7], where three different types of microstrip band stop filters based on EBGs are proposed and compared, with exagonal, octagonal, and elliptical rings.

Mobile telecommunications are another field where scattering phenomena are unavoidable and strongly contribute to the resulting electromagnetic field levels and possible coverage issues. This point may be critical if irregular geometries are involved. In [8], a model for the estimation of electromagnetic field levels in built-up areas, which enhances the COST231-Walfisch–Ikegami model [9,10], and able to effectively deal with the built-up scenarios of hilly, largely variable, and small, irregularly arranged towns, is developed and validated from an experimental viewpoint.

Finally, another interesting topic covered in this Special Issue is inverse electromagnetic scattering, with reference to the problem of microwave imaging. In particular, tomographic microwave imaging of dielectric targets inside circular conducting cylinders is discussed in [11], analyzing the performance of a nonlinear quantitative inversion technique in non-Hilbertian Lebesgue spaces. The approach studied by the authors could be exploited in many applications making use of circular enclosures, such as the case of biomedical imaging where the process requires a tight coupling of the investigated region with the surrounding background media.

Funding: This research received no external funding.

Conflicts of Interest: The authors declare no conflict of interest.

References

1. Ruello, G.; Lattanzi, R. Scattering from Spheres: A New Look into an Old Problem. *Electronics* **2021**, *10*, 216, [CrossRef]
2. Kalarickel Ramakrishnan, P.; Raffetto, M. Three-Dimensional Time-Harmonic Electromagnetic Scattering Problems from Bianisotropic Materials and Metamaterials: Reference Solutions Provided by Converging Finite Element Approximations. *Electronics* **2020**, *9*, 1065. [CrossRef]
3. Kang, Z.; Huang, M.; Li, W.; Wang, Y.; Yang, F. An Efficient Numerical Formulation for Wave Propagation in Magnetized Plasma Using PITD Method. *Electronics* **2020**, *9*, 1575. [CrossRef]
4. Fedeli, A.; Pastorino, M.; Randazzo, A. A Hybrid Asymptotic-FVTD Method for the Estimation of the Radar Cross Section of 3D Structures. *Electronics* **2019**, *8*, 1388. [CrossRef]
5. Rahmat-Samii, Y.; Mosallaei, H. Electromagnetic band-gap structures: Classification, characterization, and applications. In Proceedings of the 2001 Eleventh International Conference on Antennas and Propagation, Manchester, UK, 17–20 April 2001; Volume 2, pp. 560–564. [CrossRef]
6. Sievenpiper, D.F. High-impedance electromagnetic surfaces. Ph.D. Thesis, University of California, Los Angeles, CA, USA, 1999.
7. Zheng, X.; Jiang, T.; Lu, H.; Wang, Y. Double-Layer Microstrip Band Stop Filters Etching Periodic Ring Electromagnetic Band Gap Structures. *Electronics* **2020**, *9*, 1216. [CrossRef]
8. Schirru, L.; Ledda, F.; Lodi, M.B.; Fanti, A.; Mannaro, K.; Ortu, M.; Mazzarella, G. Electromagnetic Field Levels in Built-up Areas with an Irregular Grid of Buildings: Modeling and Integrated Software. *Electronics* **2020**, *9*, 765. [CrossRef]
9. Ikegami, F.; Yoshida, S.; Takeuchi, T.; Umehira, M. Propagation factors controlling mean field strength on urban streets. *IEEE Trans. Antennas Propag.* **1984**, *32*, 822–829. [CrossRef]
10. Walfisch, J.; Bertoni, H. A theoretical model of UHF propagation in urban environments. *IEEE Trans. Antennas Propag.* **1988**, *36*, 1788–1796. [CrossRef]
11. Fedeli, A.; Pastorino, M.; Randazzo, A.; Gragnani, G.L. Analysis of a Nonlinear Technique for Microwave Imaging of Targets Inside Conducting Cylinders. *Electronics* **2021**, *10*, 594. [CrossRef]

Article

Scattering from Spheres: A New Look into an Old Problem

Giuseppe Ruello [1,*] and Riccardo Lattanzi [2]

[1]   Department of Electrical Engineering and Information Technology, University of Napoli Federico II, 80125 Naples, Italy

[2]   Center for Advanced Imaging Innovation and Research (CAI2R) and Bernard and Irene Schwartz Center for Biomedical Imaging, Department of Radiology, New York University School of Medicine, New York, NY 10016, USA; Riccardo.Lattanzi@nyulangone.org

*   Correspondence: ruello@unina.it; Tel.: +39-081-76835-12

Abstract: In this work, we introduce a theoretical framework to describe the scattering from spheres. In our proposed framework, the total field in the outer medium is decomposed in terms of inward and outward electromagnetic fields, rather than in terms of incident and scattered fields, as in the classical Lorenz–Mie formulation. The fields are expressed as series of spherical harmonics, whose combination weights can be interpreted as reflection and transmission coefficients, which provides an intuitive understanding of the propagation and scattering phenomena. Our formulation extends the previously proposed theory of non-uniform transmission lines by introducing an expression for impedance transfer, which yields a closed-form solution for the fields inside and outside the sphere. The power transmitted in and scattered by the sphere can be also evaluated with a simple closed-form expression and related with the modulus of the reflection coefficient. We showed that our method is fully consistent with the classical Mie scattering theory. We also showed that our method can provide an intuitive physical interpretation of electromagnetic scattering in terms of impedance matching and resonances, and that it is especially useful for the case of inward traveling spherical waves generated by sources surrounding the scatterer.

Keywords: electromagnetic scattering and propagation; Mie scattering; reflection coefficient

Citation: Ruello, G.; Lattanzi, R. Scattering from Spheres: A New Look into an Old Problem. *Electronics* **2021**, *10*, 216. https://doi.org/10.3390/electronics10020216

Received: 9 September 2020
Accepted: 13 January 2021
Published: 19 January 2021

Publisher's Note: MDPI stays neutral with regard to jurisdictional claims in published maps and institutional affiliations.

1. Introduction

The study of the electromagnetic scattering from spherical objects has its origin in the work of Lorenz and Mie [1] at the turn of the 20th century. Modern formulations [2–5] of what has since been referred as "Mie scattering" arose from this outstanding work and has since been applied to light scattering, cancer detection, metamaterial theory and much more [6–15].

In Mie scattering, the total electromagnetic (EM) field outside the sphere can be expressed as the sum of the incident field, which is the field that would be there in the absence of the sphere (i.e., the scatterer), and the scattered field. The field that propagates inside the sphere is called transmitted field. The EM fields are expressed as a superposition of vector harmonics. The EM field dependence on each spherical coordinate is factorized, and the radial dependence inside the sphere is defined by means of spherical Bessel functions. To guarantee that the EM field is finite at the origin of the coordinate system, which coincides with the center of the sphere, only spherical Bessel functions of the first kind are used to describe the radial dependence inside the sphere. Outside the sphere, the EM field is typically defined by means of a combination of stationary first kind and progressive fourth kind spherical Bessel functions. Internal and external fields are linked by the continuity conditions.

Despite the elegance and compactness of the Mie analytical formulation, its physical interpretation can be challenging, because both scattered and transmitted field coefficients are expressed as a combination of Bessel and Riccati–Bessel functions. In an effort to improve

3

the physical understanding of the scattering characteristics of spheres, Debye proposed to expand each term of the Mie scattering in series [16–19]. Each term of the Debye series can be interpreted as the result of a diffraction, reflection, or transmission phenomenon at the air–sphere interface. While such formalism improves the physical comprehension, its complexity makes it practical only for a limited number of simple problems.

Building on Mie scattering theory, extensive work has been done in the last decades to provide elegant, closed-form solutions to the problem of scattering from spheres [20,21]. Such work has mainly focused on two types of scattering problems: the case of an outgoing wave incident on a spherical boundary and that of a standing wave incident on a spherical boundary. These problems, well described in [21], have been of great interest because they are useful to analyze the two important applicative fields of scattering theory: the scattering of lights and the irradiation of antennas. In the first case, the incident wave is modeled as a plane wave, whereas in the second case, the source is at the origin and the incident field is an outgoing wave. For example, Mie scattering was developed in the framework of light scattering theory, using plane waves as incident fields and adopting a number of far field approximations for the evaluation of the scattered field. This approach, however, does not help understanding complex near-field behaviors, which are important for biomedical applications.

In fact, in various diagnostic and therapeutic techniques, a local radiofrequency source is used to illuminate a body part, which can be often modeled using a dielectric sphere. Although this third type of scattering problems, in which there is an incoming spherical wave incident on a spherical boundary, is becoming increasingly relevant, current theoretical frameworks only provide limited physical intuition to analyze it. For example, in magnetic resonance imaging (MRI), where the interaction of EM fields with biological tissue affects both image quality and patient safety, dielectric spheres have been used as an approximation of the human head to simulate the performance of radiofrequency coils [22–24]. While rapid analytical approaches based on Mie scattering enable one to explore a large parameter space, they provide limited comprehension of the physical variables that govern the EM field propagation and should guide the design of MRI detectors and transmitters.

An alternative approach that could be more suitable for this type of problems is to express the scattering from the sphere in terms of equivalent transmission lines [25–27]. Schelkunoff, in particular, proposed the theory of the transmission of spherical waves [25]. This outstanding formulation introduced the basic concepts of impedance and reflection coefficient to describe the scattering from a sphere. However, its practical use has been considerably limited, mainly due to the mathematical complexity of the impedances and to the lack of a simple impedance transfer formula. In fact, the latter is critical for simplifying the description and interpretation of boundary condition problems and to relate impedances and reflection coefficients in the case of layered spheres.

This work expands the theory of equivalent transmission lines [25] by providing a closed-form solution to the problem of the scattering of an inward spherical wave on a spherical boundary, which has not been fully addressed by previous work. The main difference with respect to Mie scattering is that both the EM fields outside and inside the scatterer are expressed as a sum of inward (or incident) and outward (or reflected) waves. This field decomposition enables interpreting the ratios between field coefficients as reflection coefficients, providing an intuitive explanation of the physical phenomena that govern the EM field propagation inside the object. In addition, the resulting EM field expressions could be interpreted in terms of non-uniform equivalent transmission lines with the sphere center represented by an equivalent short load. We introduce an impedance transfer formula, which is expected to facilitate the straightforward utilization of our formalism in several applicative scenarios.

The remainder of the paper is organized as follows. In Section 2, the classical Mie scattering is recalled, whereas, in Section 3, the proposed reformulation is described. In Section 4, the two methods are compared. In Section 5, numerical examples are presented to show the advantages of the proposed method for the physical interpretation of phenomena that can be described in the framework of the Mie scattering. The main results are discussed in the concluding session.

2. Mie Scattering

The classical Mie scattering formulation is described in several research and review papers [2,4,28] which provide all the physical and mathematical details. In this section, we recall only the basic principles and the equations that will be used as a reference to introduce our approach.

From Maxwell's equations, the electric field $\mathbf{E}$ can be expressed as the solution of a Helmholtz equation:

$$\nabla^2 \mathbf{E} + k^2 \mathbf{E} = 0 \tag{1}$$

where k is the wavenumber. An identical equation can be derived for the magnetic field. In the case of scattering from spherical objects, it has been demonstrated [3,4] that the fields can be expressed in terms of the two families of vector harmonics $\mathbf{M}_{nm}$ and $\mathbf{N}_{nm}$:

$$\mathbf{M}_{nm} = B_n(kr)\left[i\pi_{nm}(\vartheta)\hat{\vartheta} - \tau_{nm}(\vartheta)\hat{\varphi}\right]e^{im\varphi} \tag{2}$$

$$\mathbf{N}_{nm} = \frac{1}{kr}\frac{d(rB_n(kr))}{dr}\left[i\pi_{nm}(\vartheta)\hat{\varphi} + \tau_{nm}(\vartheta)\hat{\vartheta}\right]e^{im\varphi} \\ + \frac{B_n(kr)}{kr}n(n+1)\mathrm{P}_n^m(\cos\vartheta)e^{im\varphi}\hat{r} \tag{3}$$

where i is the imaginary, (r, ϑ, φ) are the three coordinates of a spherical coordinate system with the origin in the center of the sphere, B_n is a spherical Bessel function of the nth order, P_n^m is the associated Legendre polynomial, $\pi_{nm}(\vartheta)$ and $\tau_{nm}(\vartheta)$ are sectorial functions, defined as:

$$\pi_{nm}(\vartheta) = m\frac{\mathrm{P}_n^m(\cos\vartheta)}{\sin\vartheta} \tag{4}$$

$$\tau_{nm}(\vartheta) = \frac{d\mathrm{P}_n^m(\cos\vartheta)}{d\vartheta} \tag{5}$$

The spherical Bessel function B_n is the solution of the Bessel equation in spherical coordinates and can be written as a combination of static first and second kind spherical Bessel functions, or traveling third and fourth kind spherical Bessel functions (also called first and second type spherical Hankel functions). In Mie scattering, the total EM field in the region outside the sphere is expressed as the sum of an incident field:

$$\mathbf{E}_i = \sum_{n=1}^{\infty}\sum_{m=-n}^{n}\left(a_{nm}\mathbf{M}_{nm}^{(1)} + b_{nm}\mathbf{N}_{nm}^{(1)}\right)$$

$$\mathbf{B}_i = \frac{k}{i\omega}\sum_{n=1}^{\infty}\sum_{m=-n}^{n}\left(a_{nm}\mathbf{N}_{nm}^{(1)} + b_{nm}\mathbf{M}_{nm}^{(1)}\right) \tag{6}$$

and a scattered field:

$$\mathbf{E}_s = \sum_{n=1}^{\infty}\sum_{m=-n}^{n}\left(c_{nm}\mathbf{M}_{nm}^{(4)} + d_{nm}\mathbf{N}_{nm}^{(4)}\right)$$

$$\mathbf{B}_s = \frac{k}{i\omega}\sum_{n=1}^{\infty}\sum_{m=-n}^{n}\left(c_{nm}\mathbf{N}_{nm}^{(4)} + d_{nm}\mathbf{M}_{nm}^{(4)}\right) \tag{7}$$

The field that instead propagates inside the sphere (i.e., the transmitted, field) is described as:

$$\mathbf{E}_t = \sum_{n=1}^{\infty}\sum_{m=-n}^{n}\left(e_{nm}\mathbf{M}_{nm}^{(1)} + f_{nm}\mathbf{N}_{nm}^{(1)}\right)$$

$$\mathbf{B}_t = \frac{k}{i\omega}\sum_{n=1}^{\infty}\sum_{m=-n}^{n}\left(e_{nm}\mathbf{N}_{nm}^{(1)} + f_{nm}\mathbf{M}_{nm}^{(1)}\right) \tag{8}$$

The superscripts $^{(1)}$ and $^{(4)}$ appended to the vector harmonics in Equations (6)–(8) indicate the kind of the spherical Bessel functions used for the description of the radial dependence. Since the incident and the transmitted fields are defined at the origin, their expressions include only spherical Bessel functions of the first kind, which impose that the field is finite at the origin. The scattered field is instead outward directed, so it is described as a superposition of spherical Bessel functions of the fourth kind (also called spherical Hankel functions of the second kind). Note that this notation is based on a $e^{i\omega t}$ time dependence. If the $e^{-i\omega t}$ notation is used, the outward waves must be described by spherical Bessel functions of the third kind.

Since the $\mathbf{M}_{nm}$ and $\mathbf{N}_{nm}$ vectors are orthogonal and, as shown by Equations (2) and (3), the $\mathbf{M}_{nm}$ vectors have no radial component, the evaluation of the field coefficients in Equations (6)–(8) can be separated in two independent problems. If all the b_{nm} coefficients are null, the electric field is orthogonal to the radial direction and the solution is called Transverse Electric (TE); if all the a_{nm} coefficients are null, the magnetic field is orthogonal to the radial direction and the solution is called Transverse Magnetic (TM).

The Mie scattering coefficients are obtained by imposing the continuity of the tangential components of the field at the sphere boundary. The complete Mie formulation is available in the literature. Here, we recall only the expressions for the scattering coefficients, which will be used as a reference to evaluate our method and can be expressed as a combination of spherical Bessel functions:

$$c_{nm} = -a_{nm}\frac{j_n'(k_1a)j_n(k_2a) - \chi j_n(k_1a)j_n'(k_2a)}{h_n^{(2)'}(k_1a)j_n(k_2a) - \chi h_n^{(2)}(k_1a)j_n'(k_2a)}$$

$$d_{nm} = -b_{nm}\frac{j_n(k_1a)j_n'(k_2a) - \chi j_n'(k_1a)j_n(k_2a)}{h_n^{(2)}(k_1a)j_n'(k_2a) - \chi h_n^{(2)'}(k_1a)j_n(k_2a)} \tag{9}$$

where a is the radius of the sphere, $\chi = \frac{k_2}{k_1}$ is the refraction index, k_1 and k_2 are the wavenumbers of the external and internal medium, respectively, and the following definition for the first derivative of the spherical Bessel functions was used:

$$B_n'(kr) = \frac{1}{kr}\frac{d(rB_n(kr))}{dr} \tag{10}$$

For a perfectly conducting sphere, the coefficients reduce to:

$$c_{nm} = -a_{nm}\frac{j_n(k_1a)}{h_n^{(2)}(k_1a)}$$

$$d_{nm} = -b_{nm}\frac{j_n'(k_1a)}{h_n^{(2)'}(k_1a)} \tag{11}$$

The above equations represent an outstanding contribution to modern physics, which inspired several works in different fields of science [1,28]. However, their physical interpretation is not immediate.

3. Proposed Method

We propose a new formulation of the scattering by spheres, in which the EM field is expressed in terms of inward and outward waves rather than incident, transmitted and scattering waves. Our method, which builds on previous work on equivalent transmission lines [20,21,25], develops from the physical consideration that the inward waves focus in the origin and the energy they carry is redistributed by the outward waves. In other words, the origin can be seen as a sink for the incoming waves and a source of outgoing waves, in accordance with the energy conservation principle.

From the mathematical point of view, in the classical Mie scattering, the energy conservation principle is respected by forcing the radial dependence of the EM field to behave as spherical Bessel functions of the first kind, guaranteeing that the fields are finite in the origin. In our approach, we keep the distinction between inward (described by Hankel functions of the first kind) and outward (described by Hankel functions of the second kind) waves and we fulfill the energy conservation principle by forcing the equality of their field coefficients inside the sphere. From the physics point of view, this constraint is nothing but an energy conservation criterion and it is coherent with the fact that inside the sphere the field behaves as a first kind spherical Bessel function, as described by the Mie scattering. In this framework, the outward waves can be viewed as the result of a reflection phenomenon happening at the origin and the scattering problem can be described by the non-uniform transmission line theory [29], with the origin acting as a perfect reflector.

In the following paragraphs, we describe the proposed formulation for the TE case. The TM solution can be calculated with an analogous procedure.

3.1. Problem Formulation

Figure 1 shows a schematic representation of the scattering problem. In Medium 1 (usually air), the total field can be expressed as a superposition of inward and outward waves (note that no distinction is made between incident and scattered field):

$$\mathbf{E}_1(r) = \sum_{n=1}^{\infty} \sum_{m=-n}^{n} E_{1nm}^{+} \mathbf{M}_{nm}^{(3)} + E_{1nm}^{-} \mathbf{M}_{nm}^{(4)}$$

$$\mathbf{H}_1(r) = \frac{k}{i\omega\mu} \sum_{n=1}^{\infty} \sum_{m=-n}^{n} E_{1nm}^{+} \mathbf{N}_{nm}^{(3)} + E_{1nm}^{-} \mathbf{M}_{nm}^{(4)} \tag{12}$$

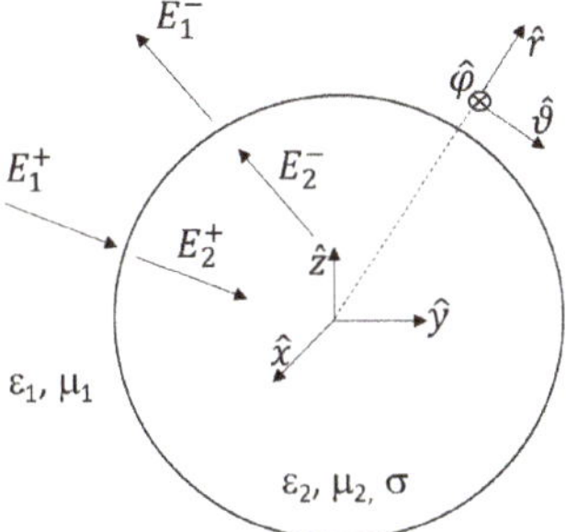

Figure 1. Geometrical representation of the scattering problem. The field in each medium is expressed as the combination of inward and outward spherical waves.

In Medium 2 (i.e., inside the sphere), the fields can also be expressed as a superposition of inward and outward waves:

$$\mathbf{E}_2(r) = \sum_{n=1}^{\infty} \sum_{m=-n}^{n} E_{2nm}^{+} \mathbf{M}_{nm}^{(3)} + E_{2nm}^{-} \mathbf{M}_{nm}^{(4)}$$

$$\mathbf{H}_2(r) = \frac{k}{i\omega\mu} \sum_{n=1}^{\infty} \sum_{m=-n}^{n} E_{2nm}^{+} \mathbf{N}_{nm}^{(3)} + E_{2nm}^{-} \mathbf{N}_{nm}^{(4)} \tag{13}$$

with the constraint that the inside the sphere inward and outward coefficients are equal ($E_{2nm}^{+} = E_{2nm}^{-}$) to ensure energy conservation.

The fields $\mathbf{E}_1(r)$ and $\mathbf{E}_2(r)$ are linked by the continuity conditions, which allow one to calculate the coefficients of the series expansion.

3.2. Characteristic Impedance

We can define an impedance term by taking the ratio between the tangential component of electric and magnetic fields. From the definition of $\mathbf{M}_{nm}$ and $\mathbf{N}_{nm}$ in Equations (2) and (3), for a single wave, the impedance can be expressed as:

$$Z_n(k_l r) = \frac{i\omega\mu}{k_l} \frac{B_n(k_l r)}{B'_n(k_l r)} \tag{14}$$

where the term $B'_n(k_l r)$ is defined in Equation (10), $l \in (1, 2)$ specifies the medium and n is the order of the Bessel function.

By observing that the ratio $\frac{\omega\mu}{k_l} = \zeta_l$ is the characteristic impedance of the l-th medium and using the logarithmic derivative of the Riccati–Bessel function:

$$\frac{B'_n(k_l r)}{B_n(k_l r)} = D_{nl} \tag{15}$$

the impedance can be written in a more compact form as $Z_n(k_l r) = i\frac{\zeta_l}{D_{nl}}$.

The impedances for the inward and outward waves are different, but their values are closely related. In fact, we can evaluate the impedance of the inward wave $Z_n^{(1)}(k_l r)$, by substituting $h_n^{(1)}(k_l r)$ in place of $B_n(k_l r)$ in Equation (14), obtaining:

$$Z_n^{(1)}(k_l r) = \frac{i\omega\mu}{k_l} \frac{h_n^{(1)}(k_l r)}{h_n^{(1)'}(k_l r)} = Z_{nl} \tag{16}$$

whereas we can evaluate the impedance of the outward wave $Z_n^{(2)}(k_l r)$ by substituting $h_n^{(2)}(k_l r)$ in place of $B_n(k_l r)$ in Equation (14), obtaining:

$$Z_n^{(2)}(k_l r) = \frac{i\omega\mu}{k_l} \frac{h_n^{(2)}(k_l r)}{h_n^{(2)'}(k_l r)} = Z_n^{(1)}(-k_l r) = \overline{Z_{nl}} \tag{17}$$

Inside the sphere, given the energy conservation constraint $E_{2nm}^{+} = E_{2nm}^{-}$, and given the identity $h_n^{(1)}(k_2 r) + h_n^{(2)}(k_2 r) = 2j_n(k_2 r)$, the total field (inward plus outward wave) behaves as a spherical Bessel function of the first kind. Therefore, by substituting $j_n(k_2 r)$ in place of $B_n(k_l r)$ in Equation (14), we can define the impedance of the total field inside the sphere as:

$$Z_n^{(J)}(k_2 r) = \frac{i\omega\mu}{k_2} \frac{j_n(k_2 r)}{j'_n(k_2 r)} = Z_{Jn2} \tag{18}$$

In Table 1, the different expressions (16)–(18) that Equation (14) can take for different Bessel functions are provided. In the last column, a compact expression that will be used in the remainder of the paper is introduced.

Table 1. Definitions of the different wave impedances.

Spherical Bessel Function	Impedance Expression	Impedance Symbol	Compact Expression
$h_n^{(1)}(k_l r)$	$\frac{i\omega\mu}{k_l} \frac{h_n^{(1)}(k_l r)}{h_n^{(1)'}(k_l r)}$	$Z_n^{(1)}(k_l r)$	Z_{nl}
$h_n^{(2)}(k_l r)$	$\frac{i\omega\mu}{k_l} \frac{h_n^{(2)}(k_l r)}{h_n^{(2)'}(k_l r)}$	$Z_n^{(2)}(k_l r)$	$\overline{Z_{nl}}$
$j_n(k_l r)$	$\frac{i\omega\mu}{k_l} \frac{j_n(k_l r)}{j'_n}$	$Z_n^{(J)}(k_l r)$	Z_{Jnl}

3.3. Field Expression: Traveling Form

For each mode of the EM field in the *l*-th medium, the tangential fields can be expressed as:

$$E_{lnm}(r) = E_{lnm}^{+} h_n^{(1)}(k_l r) + E_{lnm}^{-} h_n^{(2)}(k_l r)$$

$$
\begin{aligned}
H_{lnm}(r) &= \frac{k_l}{i\omega\mu}\left[E_{lnm}^{+} h_n^{(1)'}(k_l r) + E_{lnm}^{-} h_n^{(2)'}(k_l r) \right] \\
&= \left[\frac{E_{lnm}^{+}}{Z_{nl}} h_n^{(1)}(k_l r) + \frac{E_{lnm}^{-}}{Z_{nl}} h_n^{(2)}(k_l r) \right]
\end{aligned}
\tag{19}
$$

If we define a reflection coefficient for the electric field as:

$$\Gamma_n(k_l r) = \frac{E_l^{-} h_{nm}^{(2)}(k_l r)}{E_l^{+} h_{nm}^{(1)}(k_l r)}, \tag{20}$$

we can write the fields in a more compact form:

$$E_{lnm}(r) = E_{lnm}^{+} h_n^{(1)}(k_l r)[1 + \Gamma_n(k_l r)]$$

$$H_{lnm}(r) = \frac{E_{lnm}^{+}}{Z_{nl}} h_n^{(1)}(k_l r)\left[1 + \Gamma_n(k_l r)\frac{Z_{nl}}{Z_{nl}} \right]. \tag{21}$$

The impedance of the total field can then be expressed as a function of the reflection coefficient by taking the ratio between the electric and magnetic field in Equations (21):

$$Z_n(k_l r) = \frac{E_l(r)}{H_l(r)} = Z_{nl}\frac{1 + \Gamma_n(k_l r)}{1 + \Gamma_n(k_l r)\frac{Z_{nl}}{Z_{nl}}}. \tag{22}$$

Inside the sphere ($l = 2$), this corresponds to the impedance in Equation (18). The above expression can be easily inverted in order to express the reflection coefficients in terms of impedance as:

$$\Gamma_n(k_l r) = \frac{Z_n(k_l r) - Z_{nl}}{Z_{nl} - Z_n(k_l r)\frac{Z_{nl}}{Z_{nl}}} \tag{23}$$

The last equations provide a framework for the physical interpretation of the fields. For example, in Equations (21), the electric field is expressed as the product of the term $1 + \Gamma_n(k_l r)$, which accounts for the coherent sum of incident and reflected fields, with the term $h_n^{(1)}(k_l r)$, which accounts for the radial distribution of the energy. One advantage of the proposed approach compared to the classical Mie formulation is that the reflection coefficient and the impedance are scalar and physically interpretable engineering quantities, as opposed to the coefficients in Equation (9). For instance, impedance and reflection coefficient enable to easily evaluate, for each mode, the position of the peak of the electric field inside the sphere, providing a powerful tool for, e.g., antenna design optimization. In Section V, numerical examples are presented to illustrate applications of the proposed methods.

In the equation for the magnetic field (21), the reflection coefficient of the electric field is multiplied by the factor $Z_{nl}/\overline{Z_{nl}}$, which is a term with unitary amplitude. At the origin, the phase of this term is null; therefore, the magnetic reflection coefficient is equal to the electric reflection coefficient.

For $k_l >> n$, the phase of $Z_{nl}/\overline{Z_{nl}}$ approaches π, and the reflection coefficient of the magnetic field is the opposite of that of the electric field, as it happens for transmission lines. The phase of $Z_{nl}/\overline{Z_{nl}}$ is plotted against the sphere radius in Figure 2.

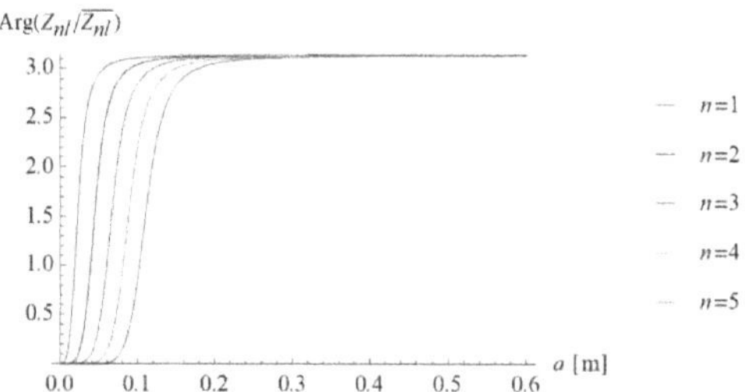

Figure 2. The phase of the term $Z_{nl}/\overline{Z_{nl}}$ is plotted against the sphere radius a. The frequency is 297.2 MHz; the sphere dielectric properties are: $\varepsilon = 50\varepsilon_0$, $\sigma = 0$; the corresponding wavenumber $k = 44.89$ m^{-1}.

3.4. Field Expression: Stationary Form

The spherical Bessel functions of the third and fourth kind can be expressed in terms of stationary spherical Bessel functions as $h_n^{(1)}(k_l r) = j_n(k_l r) + i y_n(k_l r)$ and $h_n^{(2)}(k_l r) = j_n(k_l r) - i y_n(k_l r)$. Then, the EM field can be written in stationary form as:

$$E_{lnm}(r) = \left(E_{lnm}^{+} + E_{lnm}^{-}\right) 2 j_n(k_l r) + i\left(E_{lnm}^{+} - E_{lnm}^{-}\right) 2 y_n(k_l r)$$

$$H_{lnm}(r) = \frac{k_l}{i\omega\mu}\left(E_{lnm}^{+} + E_{lnm}^{-}\right) 2 j_n'(k_l r) + i\left(E_{lnm}^{+} - E_{lnm}^{-}\right) 2 i y_n'(k_l r) \tag{24}$$

The corresponding impedance in the l-th medium can be calculated, for each mode, as the ratio between electric and magnetic fields:

$$Z_n(k_l r) = \frac{E_l(r)}{H_l(r)} = Z_{Jnl}\frac{A_{0l} + i t_{nl}}{A_{0l} + i t_{nl}'} \tag{25}$$

where

$$A_{0l} = \frac{\left(E_{lnm}^{+} + E_{lnm}^{-}\right)}{\left(E_{lnm}^{+} - E_{lnm}^{-}\right)} \tag{26}$$

$$t_{nl} = \frac{y_n(k_l r)}{j_n(k_l r)}, \tag{27}$$

and

$$t_{nl}' = \frac{y_n'(k_l r)}{j_n'(k_l r)} \tag{28}$$

Equation (25) provides a novel general expression of the impedance in any medium and it is particularly useful for the evaluation of the matching condition. For example, in case of scattering from multi-layered spheres, it provides an operative expression to evaluate the impedance in any layer as a transfer of the impedance in the origin.

As a proof of consistence with Mie scattering, we can see that, inside the sphere, given the condition $E_{2nm}^{+} = E_{2nm}^{-}$, we get that $A_{02} = \infty$ and the impedence reduces to the Z_{Jn2} in Equation (18). The importance of Equation (25) is further demonstrated by the fact that, at the origin, $Z_{Jn2} = 0$ and this allows interpreting the scattering in terms of equivalent transmission lines closed on a short circuit at the origin, as shown in Figure 3.

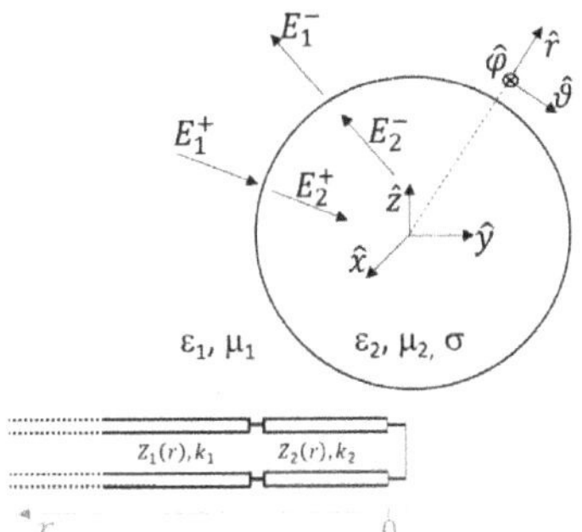

Figure 3. Equivalent transmission line for the case of an inward spherical wave incident on a spherical boundary.

3.5. Field Coefficients Evaluation

In some applications (for example, when modeling an MRI experiment) an inward wave is generated in the outer medium (by an antenna) and impinges on the sphere. Therefore, it is of interest to express the first medium outward wave (E^-_{1nm}) and the internal inward wave (E^+_{2nm}) field coefficients as a function of the first medium inward wave (E^+_{1nm}) field coefficient. To this purpose, we can use the reflection coefficient definition and the continuity of the impedance. In fact, by inverting Equation (20) and evaluating it at $r = a$, we obtain:

$$E^-_{1nm} = E^+_{1nm} \Gamma_n(k_1 a) \frac{h_n^{(1)}(k_1 a)}{h_n^{(2)}(k_1 a)} \tag{29}$$

The reflection coefficient can be written in terms of the impedance:

$$\Gamma_n(k_1 a) = \frac{Z_n(k_1 a) - Z_{n1}}{Z_{n1} - Z_n(k_1 a) R} \tag{30}$$

where R is the ratio $Z_{nl}/\overline{Z_{ni}}$, evaluated in $r = a$. By imposing the continuity condition for the impedance in $r = a$: $Z_n(k_1 a) = Z_n(k_2 a) = Z_{Jn2}(k_2 a)$, we can calculate the outward wave coefficients in terms of spherical Bessel functions as:

$$\frac{E^-_{1nm}}{E^+_{1nm}} = \frac{Z_{Jn2}(k_2 a) - Z_{n1}}{Z_{n1} - Z_{Jn2}(k_2 a) R} \cdot \frac{h_n^{(1)}(k_1 a)}{h_n^{(2)}(k_1 a)} \tag{31}$$

The previous expression is the product of two ratios. The first ratio describes a reflection coefficient, whose value depends on the dielectric properties of the media and the radius of the sphere. The second ratio represents a propagation factor, which accounts for the spatial distribution of the energy and whose value also depends on the radius of the sphere.

The coefficients of the field transmitted in the sphere (13) are evaluated by imposing the continuity of the electric field at $r = a$. From Equation (21), we have:

$$E^+_{2nm} = E^+_{1nm} \frac{h_n^{(1)}(k_1 a)}{h_n^{(1)}(k_2 a)} \frac{[1 + \Gamma_n(k_1 a)]}{[1 + \Gamma_n(k_2 a)]}. \tag{32}$$

An expression for reflection coefficient $\Gamma_n(k_2 a)$ can be obtained from Equation (20), with the physical constraint $E^+_{2nm} = E^-_{2nm}$, from being inside the sphere, and substituted in Equation (32):

$$E^+_{2nm} = E^+_{1nm} \frac{h_n^{(1)}(k_1 a)}{2 j_n(k_2 a)} [1 + \Gamma_n(k_1 a)] \tag{33}$$

The expression of the coefficient has a straightforward physical interpretation also in this case. It is the product of a propagation term that accounts for the geometry-dependent energy distribution and a transmission coefficient $\tau_n = 1 + \Gamma_n(k_1 a)$, which accounts for the energy propagated inside the sphere.

3.6. Power Evaluation

The traveling form presented in Section 3.3 is useful to describe the field inside the sphere as the coherent sum of inward and outward waves. It also allows one to evaluate the power density S_{lnm} as:

$$S_{lnm} = \frac{1}{2} E_{lnm} H^*_{lnm} \cdot \hat{r} =$$

$$= \frac{1}{2Z_{nl}} |E^+_{lnm}|^2 \left| h_n^{(1)}(k_l r) \right|^2 [1 + \Gamma_n(k_l r)][1 + \Gamma_n(k_l r) \frac{Z_{nl}}{Z_{ni}}]^* \tag{34}$$

where $\hat{r}$ is the radial unit vector defined in Figure 1.

When $kr >> n$, the magnetic field is the opposite of the electric field and the power density in each medium can be written as:

$$S_{lnm} = \frac{1}{2Z_{nl}} |E^+_{lnm}|^2 \left| h_n^{(1)}(k_l r) \right|^2 \left[1 - |\Gamma_n(k_l r)|^2 \right]. \tag{35}$$

This expression clearly shows that the inward and outward powers are decoupled. By integrating the power density on a spherical surface centered at the origin, the power dissipated inside the sphere by the n-th mode can be calculated in a straightforward manner by substituting $l = 2$ in Equation (34) and multiplying the result with $n(n+1)r^2$.

In conclusion, in Table 2, the formulas of electromagnetic fields, reflection coefficient and impedances proposed in both traveling and stationary forms are presented.

Table 2. Summary of the formulas defining the proposed framework.

	Electromagnetic Field	Reflection Coefficient	Impedance
Traveling form	$E_{lnm}(r) = E^+_{lnm} h_n^{(1)}(k_l r)[1 + \Gamma_n(k_l r)]$ $H_{lnm}(r) = \frac{E^+_{lnm}}{Z_{nl}} h_n^{(1)}(k_l r)\left[1 + \Gamma_n(k_l r)\frac{Z_{nl}}{Z_{nl}}\right]$	$\Gamma_n(k_l r) = \frac{Z_n(k_l r) - Z_{nl}}{Z_{nl} - Z_n(k_l r)\frac{Z_{nl}}{Z_{nl}}}$	$Z_n(k_l r) = Z_{nl}\frac{1 + \Gamma_n(k_l r)}{1 + \Gamma_n(k_l r)\frac{Z_{nl}}{Z_{nl}}}$
Stationary form	$E_{lnm}(r) = (E^+_{lnm} + E^-_{lnm})2j_n(k_l r) +$ $i(E^+_{lnm} - E^-_{lnm})2y_n(k_l r)$ $H_{lnm}(r) = \frac{k_l}{i\omega\mu}(E^+_{lnm} + E^-_{lnm})2j'_n(k_l r) +$ $i(E^+_{lnm} - E^-_{lnm})2iy'_n(k_l r)$	$\Gamma_n(k_l r) = \frac{Z_n(k_l r) - Z_{nl}}{Z_{nl} - Z_n(k_l r)\frac{Z_{nl}}{Z_{nl}}}$	$Z_n(k_l r) = Z_{Jnl}\frac{A_{0l} + it_{nl}}{A_{0l} + it'_{nl}}$

4. Comparison with Mie Scattering

In this section, we compare the proposed framework with the classical Mie scattering. In our approach, the electric field in the external medium E_{1nm} is expressed as the sum of inward and outward waves:

$$E_{1nm} = E^+_{1nm} h_n^{(1)}(k_1 r) + E^-_{1nm} h_n^{(2)}(k_1 r) \tag{36}$$

As both the incident and the scattered fields contribute to the outward waves, we can rewrite Equation (36) to explicitly show both contributions:

$$E_{1nm} = E^+_{1nm} h_n^{(1)}(k_1 r) + E^+_{1nm} h_n^{(2)}(k_1 r) + (E^-_{1nm} - E^+_{1nm}) h_n^{(2)}(k_1 r) \tag{37}$$

The first two terms of Equation (37) represent the incident field, and the last term is the scattering term, as defined in the Mie formulation. Therefore, considering that

$h_n^{(1)}(k_1 r) + h_n^{(2)}(k_1 r) = 2 j_n(k_1 r)$, the relationship between our coefficients and those of the classical Mie scattering is:

$$E_{1nm}^{+} = \frac{a_{nm}}{2}$$

(38)

$$E_{1nm}^{-} - E_{1nm}^{+} = c_{nm}$$

(39)

Therefore, the Mie coefficients can be easily retrieved from the coefficients of our method as:

$$\frac{c_{nm}}{a_{nm}} = \frac{1}{2}\left(\frac{E_{1nm}^{-}}{E_{1nm}^{+}} - 1\right)$$

(40)

As a further validation of our approach, in Appendix A we present the algebraic passages to obtain the c_{nm} expression presented in Equation (9) from Equation (40).

The consistence of the proposed model with Mie scattering is directly verified for the case of a perfectly conducting sphere. In fact, the continuity conditions dictate that the electric field at $r = a$ must be null. This means that the reflection coefficient of the electric field must be -1.

Therefore, the relation between the outward and inward coefficients is immediately found from Equation (29) as:

$$E_{inm}^{-} = -E_{inm}^{+} \frac{h_n^{(1)}(k_1 a)}{h_n^{(2)}(k_1 a)}$$

(41)

Substituting the last equation in Equation (40), we obtain:

$$\frac{c_{nm}}{a_{nm}} = \frac{1}{2}\left(\frac{h_n^{(1)}(k_1 a)}{h_n^{(2)}(k_1 a)} - 1\right) = -\frac{j_n(k_1 a)}{h_n^{(2)}(k_1 a)},$$

(42)

which is, in fact, the expression of the Mie scattering coefficients for the perfectly conducting sphere reported in Equation (11).

5. Numerical Results

In this section, we present examples of EM scattering from a spherical object and show how our model can provide a physically intuitive understanding of the results.

5.1. Results for the Fundamental Mode (n = 1)

We first investigated the scattering and propagation characteristics for the fundamental mode ($n = 1$) as a function of the dielectric properties of the sphere.

In all the simulations, the carrier frequency was set to 292.7 MHz, which is the operating frequency of 7 T MR scanners, the external medium was air ($\varepsilon_{r1} = 1$) and the relevant quantities were plotted for a sphere with radius a ranging from 0 to 0.6 m (corresponding to $ka = 26.93$). We investigated the dependence of the reflection coefficient (Figure 4) and impedance (imaginary and real part in Figures 5 and 6, respectively) on different values of relative electric permittivity ($\varepsilon_{r2} = 5, 25, 50$) and conductivity ($\sigma = 0, 0.05, 0.15$ and 0.5 S/m).

In addition, we plotted the radial dependence of the EM field (electric and magnetic field in Figures 7 and 8, respectively) and the power (Figures 9 and 10) inside a sphere of radius $a = 0.6$ m for the same values of relative electric permittivity ($\varepsilon_{r2} = 5, 25, 50$) and conductivity ($\sigma = 0, 0.05, 0.15$ and 0.5 S/m).

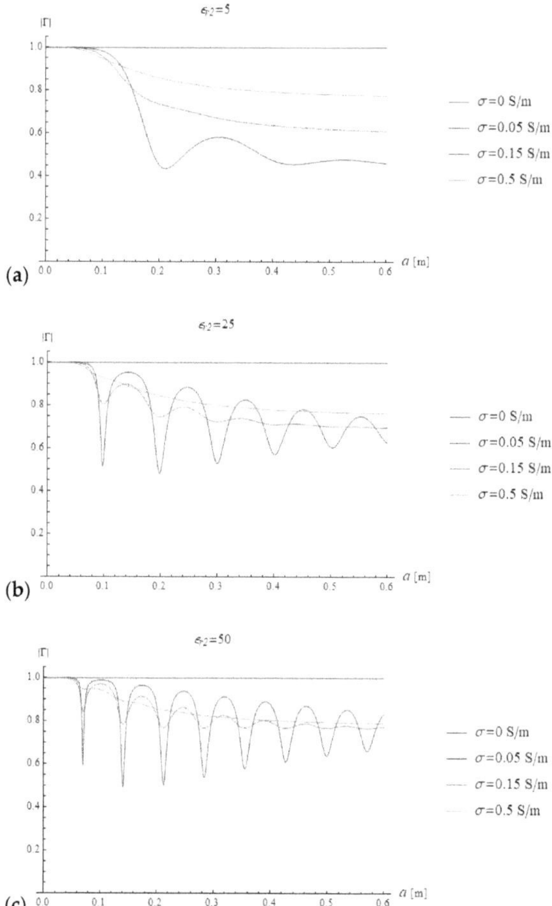

Figure 4. The modulus of the reflection coefficient is plotted as a function of the sphere radius for different values of relative permittivity ($\varepsilon_{r2} = 5$, 25 and 50 in (**a**–**c**), respectively) and conductivity ($\sigma = 0$, 0.05, 0.15 and 0.5 S/m). The frequency is set to 297.2 MHz.

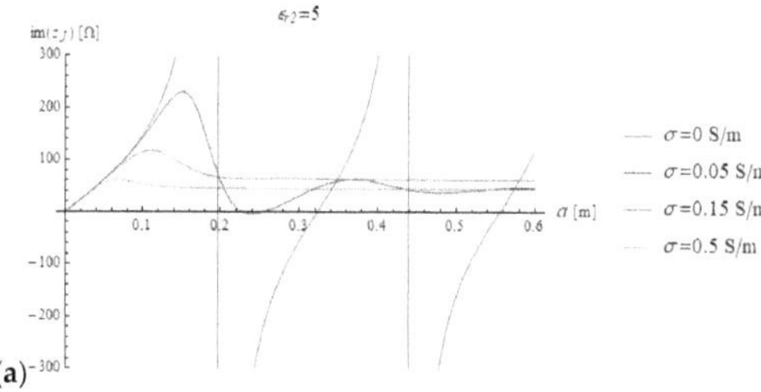

Figure 5. *Cont.*

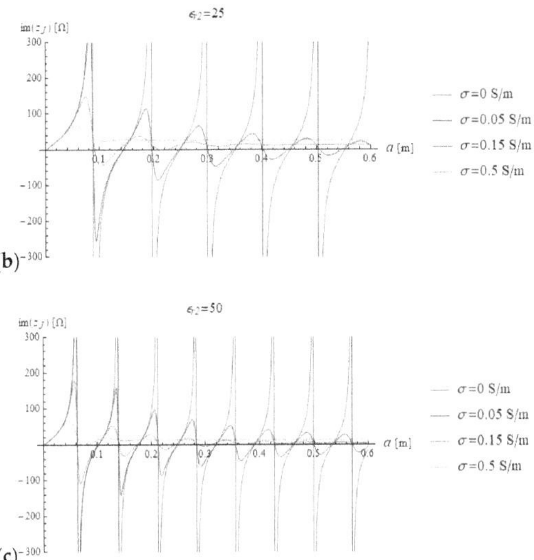

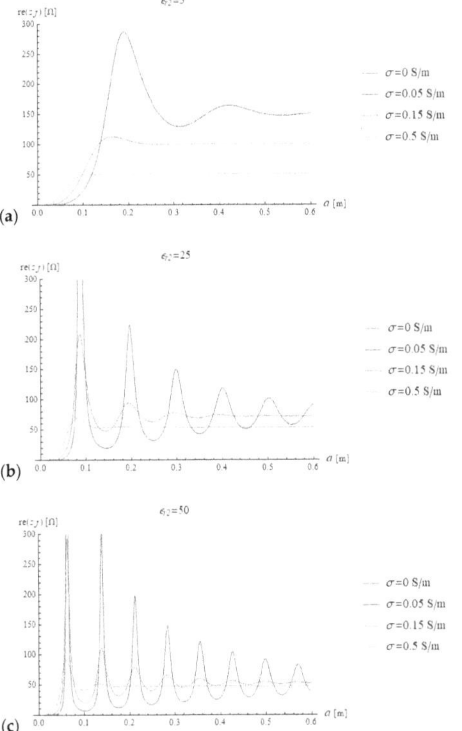

Figure 5. The imaginary part of the sphere impedance is plotted as a function of the sphere radius for different values of relative permittivity ($\varepsilon_{r2} = 5, 25$ and 50 in (**a–c**), respectively) and a range of conductivity values.

Figure 6. The real part of the sphere impedance is plotted as a function of the sphere radius for different values of relative permittivity $\varepsilon_{r2} = 5, 25$ and 50 in (**a–c**), respectively and a range of conductivity values.

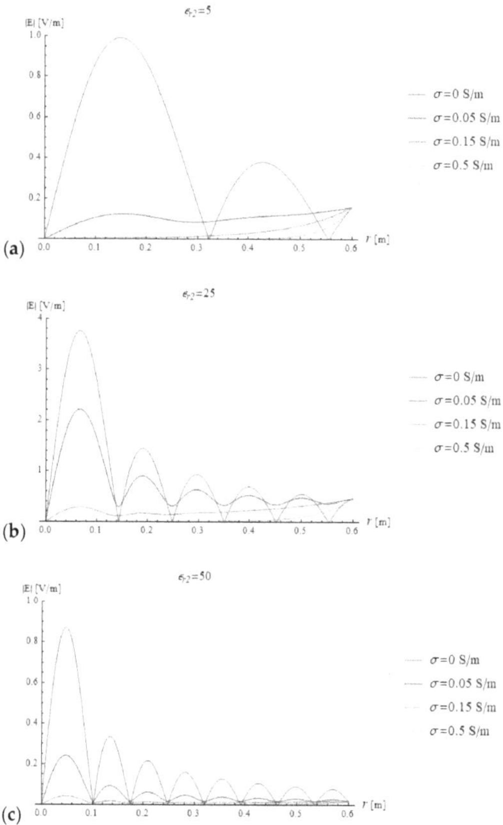

Figure 7. The amplitude of the electric field inside a sphere of radius $a = 0.6$ is plotted as a function of the radial coordinate r for different values of relative permittivity ($\varepsilon_{r2} = 5$, 25 and 50 in (**a–c**), respectively) and conductivity (see plot legend).

The results for a lossless sphere ($\sigma = 0\,\mathrm{S/m}$) are shown in blue in all plots. For a lossless sphere, Figure 4 shows that the reflection coefficient has unitary modulus, independently from the value of the dielectric constant.

This is consistent with the fact that no power is absorbed by the sphere and all the incoming power (carried by inward waves) is balanced by the outgoing power (carried by outward waves). The same result is confirmed by the fact that the sphere impedance turns out to be a pure imaginary quantity.

In the following paragraphs, we show that our approach allows one to reformulate, in terms of engineering quantities (impedances and reflection coefficients), two typical boundary-value problems described in the literature by Mie scattering: the natural oscillation modes of a sphere and the diffraction of a plane wave by a sphere [2].

It is known from the Mie scattering theory that a lossless sphere is characterized by its natural modes that obey the transcendental equations introduced in Section 9.22 of [2] (Equations (10) and (19)). With our proposed approach, the same solutions are obtained simply by imposing the continuity condition for the impedance at $r = a$: $Z_n(k_1 a) = Z_n(k_2 a)$.

By simply observing that in the inner medium there is a standing wave and that in the outer medium there is only an outward wave, it is straightforward to write the continuity condition by substituting to $Z_n(k_1 a)$ and $Z_n(k_2 a)$ the impedance expressions reported in the second and third rows of Table 1, respectively.

As a result, the condition can be simply written as:

$$\overline{Z_{nI}} = Z_{Jn2} \tag{43}$$

or substituting the explicit expression reported in the second column of Table 1 as:

$$\frac{1}{k_1}\frac{h_n^{(2)}(k_1 a)}{h_n^{(2)'}(k_1 a)} = \frac{1}{k_2}\frac{j_n(k_2 a)}{j_n'(k_2 a)} \tag{44}$$

Equation (43) provides the transcendental equation introduced in Section 9.22 of [2], whose solutions provide the natural frequencies of the sphere, demonstrating that a set of natural modes exists.

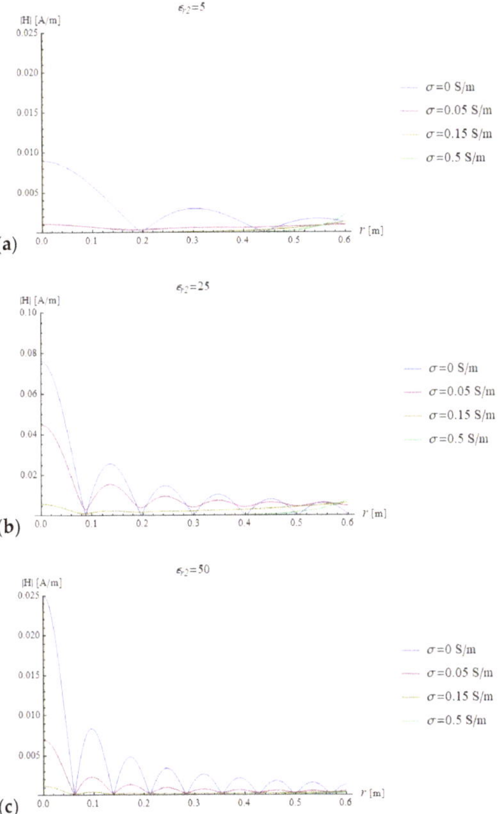

Figure 8. The amplitude of the magnetic field inside a sphere of radius $a = 0.6$ is plotted as a function of the radial coordinate r for different values of relative permittivity ($\varepsilon_{r2} = 5$, 25 and 50 in (**a–c**), respectively) and conductivity (see plot legend).

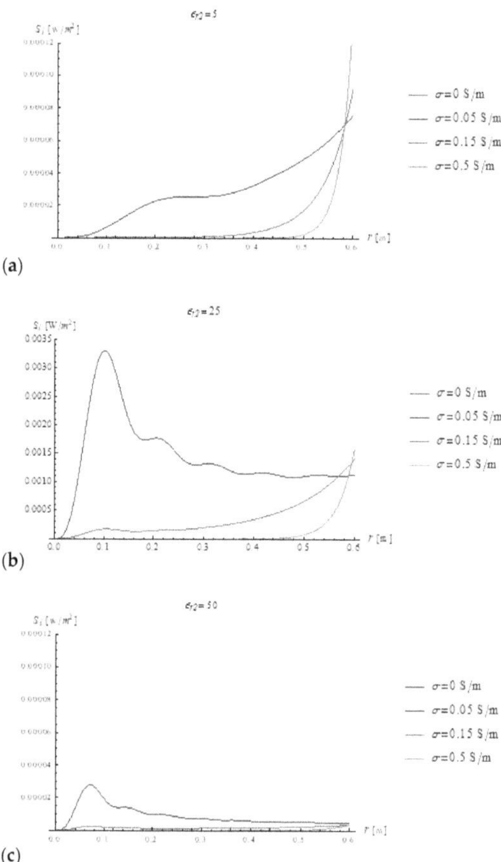

Figure 9. The amplitude is plotted as a function of the radial coordinate for different values of relative permittivity ($\varepsilon_{r2} = 5$, 25 and 50 in (**a**–**c**), respectively) and conductivity (see plot legend).

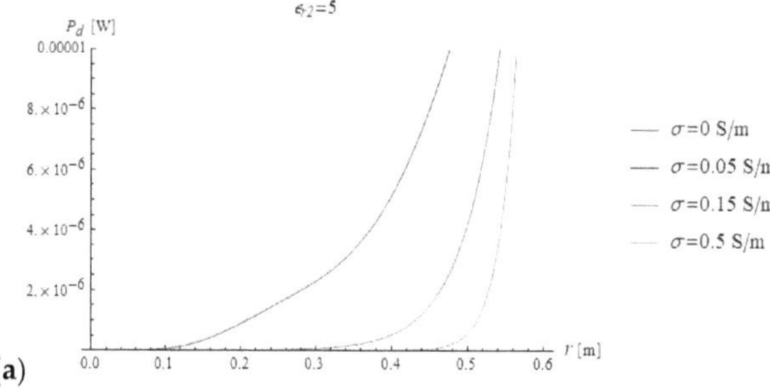

Figure 10. *Cont.*

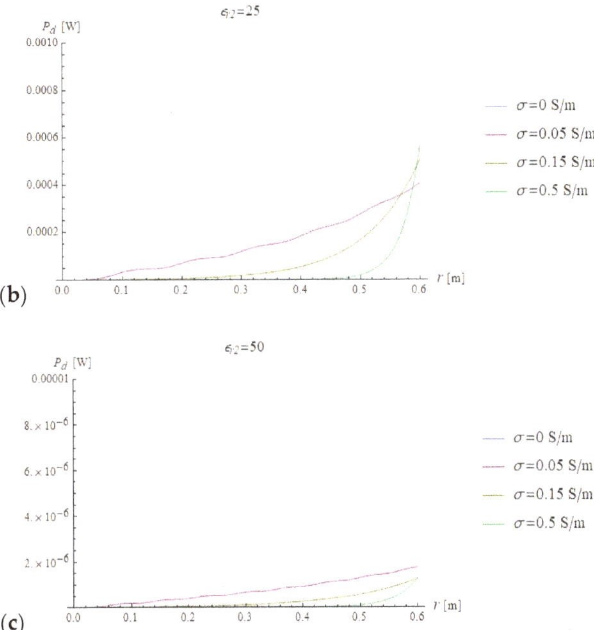

Figure 10. The power dissipated inside a sphere is plotted as a function of the radial coordinate for different values of relative permittivity ($\varepsilon_{r2} = 5, 25$ and 50 in (**a**–**c**), respectively) and conductivity (see plot legend).

The main advantages of using our formulation are:

1. By employing the concept of impedance, we derived Equation (44) by means of simple and easily interpretable physical arguments.
2. Equation (44) can be easily extended to the case of a stratified sphere, which represents a considerable simplification compared to the classical approach that requires a complete reformulation of the scattering problem for each layer.

The diffraction of EM energy from a sphere can also be analyzed using our proposed framework. It is known that, when a plane wave impinges on a homogeneous lossless dielectric sphere, the EM field transmitted in the sphere (whose analytical expression is provided in Equation (33)) has a resonance-like behavior, governed by the geometric and dielectric characteristics of the sphere. From Equation (33), we can easily find the resonance condition by maximizing the ratio between the inward field coefficients of the two media, which occurs when the $1 + \Gamma_n(k_1 a)$ factor reaches its maximum value, i.e., when the value of $\Gamma_n(k_1 a)$ is real and positive. Therefore, our formulation intuitively shows that resonances occur when outward and inward waves have electric fields in phase and magnetic fields in antiphase on the sphere surface. This means that the magnetic field is null on the sphere surface, which is consistent with the fact that the impedance $Z_{Jn2}(k_2 a)$ is infinite in resonance condition.

In perfect analogy with uniform transmission lines, our formulation allows one to interpret the resonance phenomenon in terms of the sphere's impedance and, therefore, in terms of the phase difference between electric and magnetic fields on the spherical interface, which is responsible for the field distribution outside and inside the sphere and for the consequent energy storage.

The previous phenomena can be appreciated also directly from the plots of the impedance and the EM field. For example, a resonance effect can be seen in Figures 7b and 8b, where the maximum values of the amplitude of the electric and magnetic fields for $\varepsilon_{r2} = 25$ are considerably higher than the corresponding values for $\varepsilon_{r2} = 5$ (Figures 7a and 8a) and $\varepsilon_{r2} = 50$ (Figures 7c and 8c). For the fundamental mode ($n = 1$), for $a = 0.6$ m and $\varepsilon_{r2} = 25$, the product $ka = 18.67$ almost coincides with the sixth null of the derivative of the spherical Bessel function (18.79). From Figure 5 (see above) and Equation (31), we know that this corresponds to infinite impedance at the sphere surface, which means that the amplitude of the $1 + \Gamma_n(k_1 a)$ factor is maximum and, therefore, the mode resonates.

In a similar manner, also in the TM case, we can obtain discrete resonance values corresponding to infinite values of the admittance at the interface, by looking at the nulls of the impedance at the interface.

For the case of a lossy sphere, the field is attenuated during its propagation and the inward–outward interferences are damped.

It is interesting to note that in this case the real part of the impedance is non-null, with peaks in correspondence of the resonant frequencies. In fact, the sphere impedance approaches the intrinsic impedance of the external medium, resulting in a minimum of the reflection coefficient (see Equation (31)).

Using our proposed framework, phenomena like those described above can be interpreted with traditional engineering concepts, such as an impedance pseudo-matching (in fact, perfect matching is not possible for a single sphere with real frequencies).

Relevant insight can also be gained from the power density plots. In a lossless sphere, the net incoming power density is fully balanced by the outgoing one and the net flux is null for all permittivity values (see Figures 9 and 10, where the blue line is superimposed to the x axis). This phenomenon is also shown by Equations (34) and (35), where it is clear that the net power density is null if the magnitude of the reflection coefficient $|\Gamma_n(k_l r)|$ is unitary.

In the presence of losses, the power density closely follows the square of the amplitude of the electric field, and when $\varepsilon_{r2} = 25$, which is the in-resonance condition for the fundamental mode, for low losses ($\sigma = 0.05$ S/m), the power density is also large in proximity of the origin. If the losses are significant ($\sigma = 0.15$ and $\sigma = 0.5$ S/m), the power is mostly dissipated by the peripheral region of the sphere and the power density inside the sphere is almost null.

Figure 10 shows the power dissipation, which is the integral of the power density and is a monotonic function of the radial coordinate r. As expected, the higher the conductivity, the lower the capacity of the field to penetrate inside the sphere, and the power is mainly dissipated in the peripheral region of the sphere.

The power value at the boundary of the sphere (i.e., for $a = 0.6$ m in Figure 10) represents the total dissipated power inside the sphere.

It is worth highlighting that, with our formalism, the power was evaluated with a simple scalar expression (Equation (34)), without solving integrals, which represents an advantage compared to Mie scattering, in terms of the computation time and complexity associated with the integration of rapidly oscillating functions.

5.2. Full Modal Analysis

The behavior of the EM field for modes characterized by the higher order of the spherical Bessel functions was first investigated for the case of a lossless sphere.

Figures 11 and 12 show the transmitted electric and magnetic field as a function of the radial coordinate in a sphere of radius $a = 0.6$ m, evaluated for the first four modes of the Bessel functions.

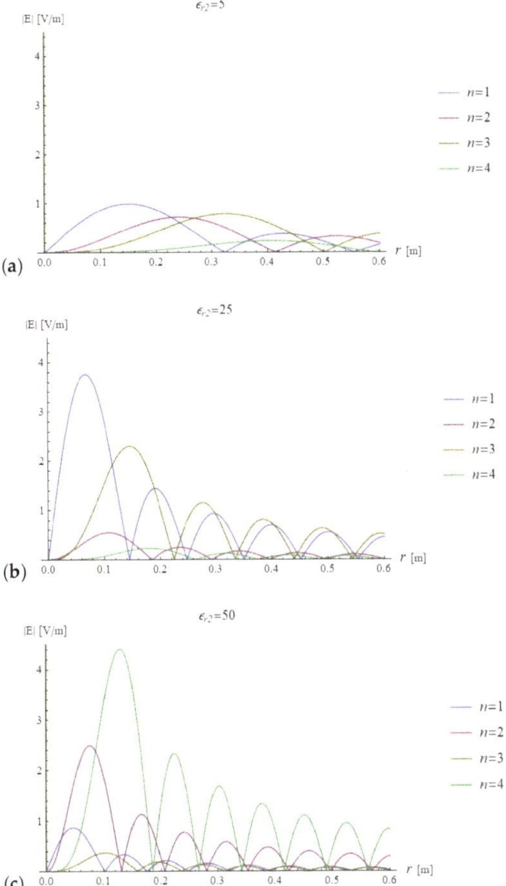

Figure 11. Electric field distribution as a function of the radial coordinate in a lossless ($\sigma = 0\,\mathrm{S/m}$) sphere of radius $a = 0.6$ m for the first four modes. Results are plotted for different values of relative permittivity ($\varepsilon_{r2} = 5, 25$ and 50 in (**a**–**c**), respectively).

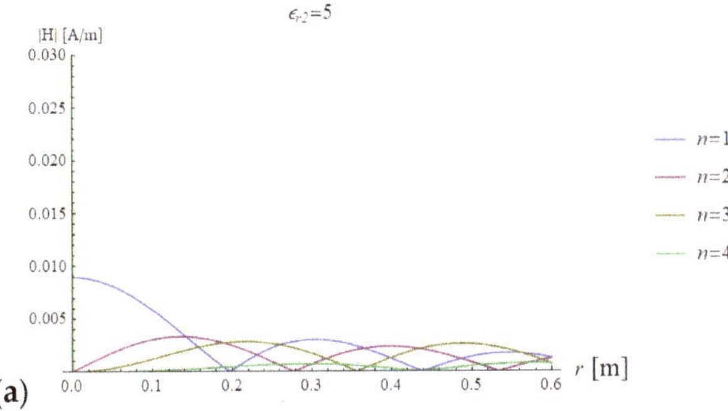

Figure 12. *Cont.*

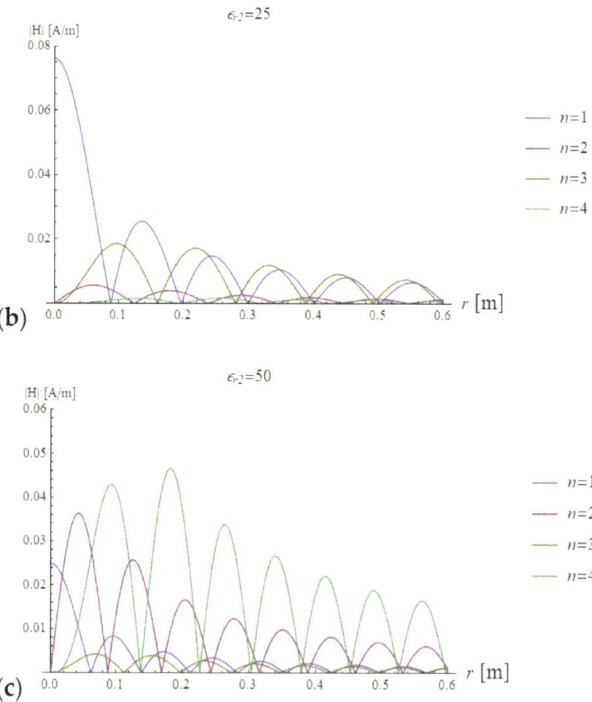

Figure 12. Magnetic field distribution as a function of the radial coordinate in a lossless ($\sigma = 0 \, \mathrm{S/m}$) sphere of radius $a = 0.6$ m for the first four modes. Results are shown for different values of relative permittivity ($\varepsilon_{r2} = 5, 25$ and 50 in (**a–c**), respectively).

For a low dielectric contrast ($\varepsilon_{r2} = 5$), none of the modes is resonant and the peaks of the fields are slightly decreasing with the order of the Bessel function. At the origin, the electric field is always null, due to field symmetry, and the only mode contributing to the magnetic field is the first mode.

Figure 11b,c and Figure 12b,c show that the field distribution is strongly influenced by the resonance conditions.

For example, as already noted when describing Figures 7b and 8b, when $\varepsilon_{r2} = 25$ and $a = 0.6$ m, the first mode is close to the resonance condition and, therefore, the corresponding electric and magnetic amplitudes both have a peak (Figures 11b and 12b). In addition, in the same figures, the third mode is also amplified, and this is due to the fact that $ka = 18.67$ is close to the fifth null of the derivative of the third order Bessel function, which corresponds to an infinite impedance and, therefore, to the resonance condition.

In Figures 11c and 12c, a similar pattern occurs for the second and fourth modes, which are on resonance when $\varepsilon_{r2} = 50$ and $a = 0.6$ m, which means they are the main contributors (among the first 4 modes analyzed in the plots) to the electric and magnetic fields.

A possible application of these results would be in the optimization of computation time. In fact, using the interpretation provided above, it is straightforward to order the modes based on their contribution to either the electric or magnetic field, and perform calculations only for the limited set of modes that contributes the most. In addition, the proposed scattering formulation in terms of engineering quantities could be useful for design optimization. For example, in MRI, it could guide the design of novel materials to be integrated with the radiofrequency coils ([30]) that can force impedance matching only for specific modes, in order to maximize the magnetic field (i.e., the source of the MRI signal) in a specific region while limiting power deposition (i.e., losses) over the entire sample [24].

In Figure 13, we show the EM distribution in the case of a lossy sphere, with conductivity $\sigma = 0.05$ S/m. As expected, all the modes are attenuated by the lossy medium. In addition, the resonance phenomenon cannot take place as for the case of a lossless sphere, because here the impedance cannot be infinite (see Figure 4), which results in considerable amplitude damping of the resonant modes (see y-axis scale of Figure 12 vs. Figures 10 and 11).

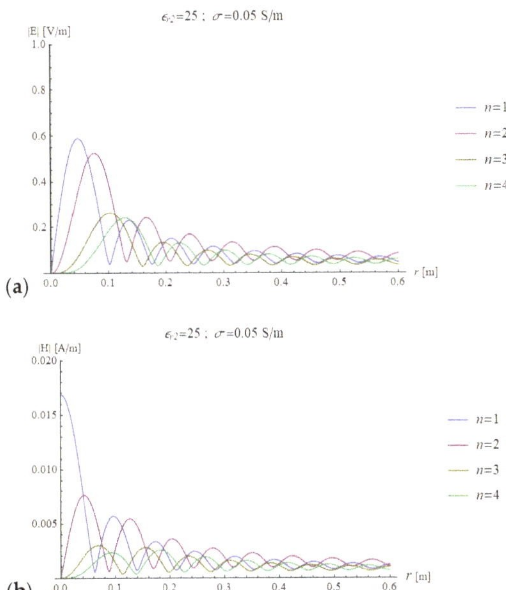

Figure 13. Electric (**a**) and magnetic (**b**) field distribution as a function of the radial coordinate for the first four modes. The sphere has a radius of $a = 0.6$ m, conductivity of 0.05 S/m and a relative permittivity of 50, mimicking the electrical properties of average brain tissue at 297.2 MHz.

6. Discussion and Conclusions

In this work, we presented a theoretical framework to study the electromagnetic scattering and propagation characteristics in spherical objects. We demonstrated the proposed approach for the simple case of a homogeneous sphere, showing that it is fully consistent with the established Mie scattering theory. Our formulation extends previous work on equivalent transmission lines and the main advantage over Mie scattering is the possibility to analyze the propagation of the EM field in terms of reflection and transmission coefficients, making the physical understanding of the field distribution more intuitive. This goal was achieved by describing the fields inside and outside the sphere as a superposition of inward and outward waves, and forcing the equality of the inward and outward field coefficients inside the sphere to respect the energy conservation principle.

The described approach can be directly applied to research problems that currently use Mie scattering. One example is the design of metasurfaces, which enable unconventional phenomena, such as perfect absorption, holography, electromagnetic invisibility and much more [10,31–33]. In such application, the Mie coefficients are combined with homogenization techniques to evaluate the electromagnetic response of an array of high permittivity dielectric spheres, deriving a surface impedance. A possible implication of this study could be the expression of the surface impedance in terms of the impedance of the individual dielectric spheres, with an improvement of the physical interpretation behind the surface design.

Our model could be generalized in a straightforward manner to describe the scattering by multi-layered spheres, which has applications in several fields. In fact, our formulation can be seen as an extension of the theory of spherical transmission lines [25]. In particular, while the concepts of impedance and reflection coefficients in the analysis of the scattering from spheres were previously described, the significant novelty of this work is the introduction of an intuitive impedance transfer formula that simplifies the definition of the boundary conditions between layers. A possible application of our comprehensive theoretical framework could be the optimization of the properties of high-permittivity coil substrates that are used to manipulate the EM field distribution to improve the diagnostic performance of MRI coils [24,30].

While the proposed formulation provides intuitive physical insight if the sources surround the spherical object, as in various biomedical applications, it is also applicable to the classical problem of scattering of plane waves from spheres. In fact, the formulation is valid for any source that can be expressed as a linear combination of spherical waves.

One limitation of our approach is its effectiveness when a large number of modes is needed to describe the total electromagnetic field. In fact, in such case, it could be challenging to intuitively grasp an overall physical interpretation of the scattering from the analysis of the individual modes. Nevertheless, the framework would still allow one to identify a few dominant modes for the case of interest and study their behavior first.

In conclusion, the proposed method allows for expressing the scattering from spheres in terms of relevant engineering entities, providing physicists and engineers with a new tool to interpret the Mie scattering mathematical results, and to design systems that involve spherical scatterers with a full physical comprehension of the underlying phenomena.

Author Contributions: The authors have read and agreed to the published version of the manuscript. Conceptualization and formal analysis, R.L. and G.R.; Methodology, G.R.; Writing–original draft, G.R.; Writing–review & editing, R.L. and G.R. All authors have read and agreed to the published version of the manuscript.

Funding: This work was supported in part by the U.S.—Italy Fulbright Commission, by awards NIH R01 EB024536 and NSF 1453675. It was performed under the rubric of the Center for Advanced Imaging Innovation and Research (CAI^{2}R, www.cai2r.net), a NIBIB Biomedical Technology Resource Center (NIH P41 EB017183).

Conflicts of Interest: The authors declare no conflict of interest.

Appendix A

In this appendix, we present the algebraic steps necessary to obtain the classical Mie scattering formulation from the reflection coefficient introduced in this paper. By using Equation (18) in Equation (31), we can write the ratio between outward and inward coefficients in Medium 1 as:

$$\frac{E_{1nm}^{-}}{E_{1nm}^{+}} = \frac{\frac{1}{k_2}\frac{j_n(k_2 a)}{j_n'(k_2 a)} - \frac{1}{k_1}\frac{h_n^{(1)}(k_1 a)}{h_n^{(1)'}(k_1 a)}}{\frac{1}{k_1}\frac{h_n^{(1)}(k_1 a)}{h_n^{(1)'}(k_1 a)} - \frac{1}{k_2}\frac{j_n(k_2 a)}{j_n'(k_2 a)}R}\frac{h_n^{(1)}(k_1 a)}{h_n^{(2)}(k_1 a)} = \frac{k_1 h_n^{(1)'}(k_1 a)j_n(k_2 a) - k_2 j_n'(k_{2a})h_n^{(1)}(k_1 a)}{k_2 j_n'(k_{2a})h_n^{(1)}(k_1 a) - k_1 h_n^{(1)'}(k_1 a)j_n(k_2 a)R}\frac{h_n^{(1)}(k_1 a)}{h_n^{(2)}(k_1 a)} \tag{A1}$$

where the term R is defined as:

$$R = \frac{h_n^{(1)}(k_1 a)}{h_n^{(2)}(k_1 a)}\frac{h_n^{(2)'}(k_1 a)}{h_n^{(1)'}(k_1 a)} \tag{A2}$$

Substituting Equation (A2) in Equation (A1), we obtain:

$$\frac{E_{1nm}^{-}}{E_{1nm}^{+}} = \frac{h_n^{(1)'}(k_1a)j_n(k_2a) - \chi j_n'(k_{2a})h_n^{(1)}(k_1a)}{\chi j_n'(k_2a)h^{(2)}(k_1a) - h_n^{(2)'}(k_1a)j_n(k_2a)} \tag{A3}$$

Using simple algebra, we can then derive the following quantity:

$$\frac{1}{2}\left(\frac{E_{1nm}^{-}}{E_{1nm}^{+}} - 1\right) = \frac{j_n'(k_1a)j_n(k_2a) - \chi j_n'(k_2a)j_n(k_1a)}{\chi j_n'(k_2a)h^{(2)}(k_1a) - \overset{\cdot}{h}^{(2)}(k_1a)j_n(k_2a)} \tag{A4}$$

The left side of Equation (A4) is the ratio between scattered and incident fields expressed with the proposed formalism. The right side of the equation shows that this ratio has the same analytical expression of the TE Mie scattering coefficients presented in Equation (9), confirming that our proposed method is equivalent to the Mie approach.

References

1. Mie, G. Contributions to the Optics of Turbid Media: Particularly of Colloidal Metal Solutions. *Ann. Phys.* **1908**, *3*, 377–445. [CrossRef]
2. Stratton, J.A. *Electromagnetic Theory*; McGraw-Hill: New York, NY, USA, 1941.
3. Bohren, C.F. *Absorption and Scattering of Light by Small Particles*; WILEY-VCH: Weinheim, Germany, 1983; ISBN 9783527406647.
4. Frezza, F.; Mangini, F.; Tedeschi, N. Introduction to electromagnetic scattering: Tutorial. *J. Opt. Soc. Am. A* **2018**, *35*, 163. [CrossRef]
5. Tsang, L.; Kong, J.A.; Ding, K.H. *Scattering of Electromagnetic Waves, Vol. 1: Theory and Applications*; Wiley Interscience: Hoboken, NJ, USA, 2000.
6. Wirdatmadja, S.; Johari, P.; Desai, A.; Bae, Y.; Stachowiak, E.; Stachowiak, M.; Jornet, J.M.; Balasubramaniam, S. Analysis of Light Propagation on Physiological Properties of Neurons for Nanoscale Optogenetics. *IEEE Trans. Neural. Syst. Rehabil. Eng.* **2019**, *27*, 108–117. [CrossRef]
7. Sudiarta, I.W.; Chýlek, P. Mie scattering by a spherical particle in an absorbing medium. *Appl. Opt.* **2002**, *41*, 3545. [CrossRef]
8. Mikulski, J.J.; Murphy, E.L. The Computation of Electromagnetic Scattering from Concentric Spherical Structures. *IEEE Trans. Antennas Propag.* **1963**, *11*, 169–177. [CrossRef]
9. Norouzian, F.; Marchetti, E.; Gashinova, M.; Hoare, E.; Constantinou, C.; Gardner, P.; Cherniakov, M. Rain Attenuation at Millimetre Wave and Low-THz Frequencies. *IEEE Trans. Antennas Propag.* **2019**, *68*, 421–431. [CrossRef]
10. Younesiraad, H.; Bemani, M.; Nikmehr, S. Scattering suppression and cloak for electrically large objects using cylindrical metasurface based on monolayer and multilayer mantle cloak approach. *IET Microw. Antennas Propag.* **2019**, *13*, 278–285. [CrossRef]
11. Prophete, C.; Sik, H.; Kling, E.; Carminati, R.; De Rosny, J. Terahertz and Visible Probing of Particles Suspended in Air. *IEEE Trans. Terahertz Sci. Technol.* **2019**, *9*, 120–125. [CrossRef]
12. Naqvi, Z.; Green, M.; Smith, K.; Wang, C.; Del'Haye, P.; Her, T.H. Uniform thin films on optical fibers by plasma-enhanced chemical vapor deposition: Fabrication, mie scattering characterization, and application to microresonators. *J. Light. Technol.* **2018**, *36*, 5580–5586. [CrossRef]
13. Figueiredo, R.J.; Backman, V.; Liu, Y.; Paladugula, J. Architecture and performance of a grid-enabled lookup-based biomedical optimization application: Light scattering spectroscopy. *IEEE Trans. Inf. Technol. Biomed.* **2007**, *11*, 170–178. [CrossRef] [PubMed]
14. Smith, D.D.; Fuller, K.A. Photonic bandgaps in Mie scattering by concentrically stratified spheres. *J. Opt. Soc. Am. B* **2002**, *19*, 2449. [CrossRef]
15. Chen, H.; Wu, B.I.; Zhang, B.; Kong, J.A. Electromagnetic wave interactions with a metamaterial cloak. *Phys. Rev. Lett.* **2007**, *99*. [CrossRef] [PubMed]
16. Debye, P. Der Lichtdruck auf Kugeln von Beliebigem Material. *Ann. Phys.* **1909**, *335*, 57. [CrossRef]
17. Hovenac, E.A.; Lock, J.A. Assessing the contributions of surface waves and complex rays to far-field Mie scattering by use of the Debye series. *J. Opt. Soc. Am. A* **1992**, *9*, 781. [CrossRef]
18. Li, R.; Han, X. Debye Series Analysis of Forward Scattering by a Multi-layered Sphere. *PIERS Online* **2007**, *3*, 209–212. [CrossRef]
19. Li, R.; Han, X.; Jiang, H.; Ren, K.F. Debye series for light scattering by a multilayered sphere. *Appl. Opt.* **2006**, *45*, 1260–1270. [CrossRef]
20. Harrington, R.F. *Time-Harmonic Electromagnetic Fields*; McGraw-Hill: New York, NY, USA, 1961.
21. Chew, W.C. *Waves and Fields in Inhomogeneous Media*; IEEE: Piscataway, NJ, USA, 1995.
22. Pfrommer, A.; Henning, A. On the Contribution of Curl-Free Current Patterns to the Ultimate Intrinsic Signal-to-Noise Ratio at Ultra-High Field Strength. *NMR Biomed.* **2017**, *30*, 1–16. [CrossRef]
23. Lee, H.H.; Sodickson, D.K.; Lattanzi, R. An analytic expression for the ultimate intrinsic SNR in a uniform sphere. *Magn. Reson. Med.* **2018**, *80*, 2256–2266. [CrossRef]
24. Vaidya, M.V.; Sodickson, D.K.; Collins, C.M.; Lattanzi, R. Disentangling the effects of high permittivity materials on signal optimization and sample noise reduction via ideal current patterns. *Magn. Reson. Med.* **2019**, *81*, 2746–2758. [CrossRef]
25. Schelkunoff, S.A. Transmission Theory of Spherical Waves. *Electr. Eng.* **1938**, *57*, 744–750. [CrossRef]

26. Wait, J.R. Electromagnetic scattering from a radially inhomogeneous sphere. *Appl. Sci. Res.* **1962**, *10*, 441–450. [CrossRef]
27. Panaretos, A.H.; Werner, D.H. Transmission line approach to quantifying the resonance and transparency properties of electrically small layered plasmonic nanoparticles. *J. Opt. Soc. Am. B* **2014**, *31*, 1573–1580. [CrossRef]
28. Wriedt, T. Mie Theory: A Review. In *The Mie Theory*; Springer: Berlin/Heidelberg, Germany, 2012; pp. 53–71, ISBN 9783642282515.
29. Latmiral, G.; Franceschetti, G.; Vinciguerra, R. Analysis and Synthesis of Nonuniform Transmission Lines or Stratified Layers. *J. Res. Natl. Bur. Stand. D Radio Propag.* **1963**, *67D*, 331. [CrossRef]
30. Webb, A.G. Dielectric Materials in Magnetic Resonance. *Concepts Magn. Reson. Part A* **2011**, *38A*, 148–184. [CrossRef]
31. Monti, A.; Alù, A.; Toscano, A.; Bilotti, F. Surface Impedance Modeling of All-Dielectric Metasurfaces. *Trans. Antennas Prop.* **2019**. [CrossRef]
32. Soric, J.C.; Fleury, R.; Monti, A.; Toscano, A.; Bilotti, F.; Alù, A. Controlling scattering and absorption with metamaterial covers. *IEEE Trans. Antennas Propag.* **2014**, *62*, 4220–4229. [CrossRef]
33. Kruk, S.; Kivshar, Y. Functional Meta-Optics and Nanophotonics Govern by Mie Resonances. *ACS Photonics* **2017**, *4*, 2638–2649. [CrossRef]

Article

Three-Dimensional Time-Harmonic Electromagnetic Scattering Problems from Bianisotropic Materials and Metamaterials: Reference Solutions Provided by Converging Finite Element Approximations

Praveen Kalarickel Ramakrishnan [†] and Mirco Raffetto [*,†]

Department of Electrical, Electronic, Telecommunications Engineering and Naval Architecture,
University of Genoa, Via Opera Pia 11a, I–16145 Genoa, Italy; pravin.nitc@gmail.com
* Correspondence: mirco.raffetto@unige.it; Tel.: +39-010-3352796
† These authors contributed equally to this work.

Received: 29 May 2020; Accepted: 24 June 2020; Published: 29 June 2020

Abstract: A recently developed theory is applied to deduce the well posedness and the finite element approximability of time-harmonic electromagnetic scattering problems involving bianisotropic media in free-space or inside waveguides. In particular, three example problems are considered of which one deals with scattering from plasmonic gratings that exhibit bianisotropy while the other two deal with bianisotropic obstacles inside waveguides. The hypotheses that guarantee the reliability of the numerical results are verified, and the ranges of the constitutive parameters of the media involved for which the finite element solutions are guaranteed to be reliable are deduced. It is shown that, within these ranges, there can be significant bianisotropic effects for the practical media considered as examples. The ensured reliability of the obtained results can make them useful as benchmarks for other numerical approaches. To the best of our knowledge, no other tool can guarantee reliable solutions.

Keywords: electromagnetic scattering; time-harmonic electromagnetic fields; bianisotropic media; metamaterials; variational formulation; well posedness; finite element method; convergence of the approximation

---

1. Introduction

Bianisotropic media have important applications in numerous practical problems ranging from the microwave to photonic frequency bands [1–4]. The electromagnetic problems involving such media admit analytical solutions only in very specialized cases, and numerical simulators are necessary to solve the vast majority of them.

In this context, the reliability of the numerical solvers is of utmost importance, and results guaranteeing the well posedness of the problems and the convergence of the numerical solutions are crucial. Some of the previous papers that addressed this issue were limited in their applications [5–8]. For instance, the work in [5] made strong assumptions on the losses to guarantee the reliability of the results. The results in [6] were derived for two-dimensional problems involving axially moving cylinders, whereas [7] dealt only with evolution problems inside cavities. As for [8], the constitutive parameters were taken to be smooth, which did not allow applications to radiation and scattering problems.

A considerable generalization of the conditions that allow ensuring the well posedness of the problems and the convergence of the finite element solutions was recently achieved in [9]. A set of non-restrictive hypotheses was shown to guarantee such results for three-dimensional time-harmonic problems involving bianisotropic media. The authors applied the theory to rotating axisymmetric objects, where the effect of motion induced bianisotropy. However, the theoretical results derived were applicable to a much wider range of bianisotropic materials and metamaterials.

In this paper, we exploit the recently developed theory to obtain novel results. We consider examples of practical problems discussed in the open literature for which, to the best of our knowledge, none of the previous theories were able to guarantee the reliability of the numerical results. In particular, we study the electromagnetic scattering from plasmonic gratings, which exhibits bianisotropy [1], and from bianisotropic obstacles in waveguides [10,11]. We demonstrate the application of the theory in [9] to derive the conditions on the constitutive parameters of these problems that guarantee the reliability of the results. The numerical solutions of the problems are calculated under such conditions, which, owing to the reliability assured by the theory, can be used as references for other numerical solvers. As far as we are aware, no other tool is available that allows obtaining benchmark solutions for these problems.

The paper is organized as follows. In Section 2, the problem is defined and the theory is summarized to guide the reader in its application. Section 3 contains the main results of the paper, where the application of the theory is demonstrated and reliable solutions are obtained. Section 4 provides the conclusions.

2. Mathematical Description of the Problem

In this paper, we are interested in electromagnetic problems that involve bianisotropic media under time-harmonic excitation, which were studied in [9]. While the full details of the problem definition and results are available in the reference, here we provide a summary of the main points in order to ease the understanding of the present developments.

The problem is formulated in an open, bounded, and connected domain $\Omega \in \mathbb{R}^3$, which has a Lipschitz continuous stationary boundary denoted by Γ. To take into account electromagnetic problems involving inhomogeneous materials, we assume that Ω can be decomposed into m subdomains (see HD3 of [9]) denoted $\Omega_i, i \in I = \{1, ..., m\}$.

The time-harmonic sources imply that all the resulting fields are in turn time-harmonic and the assumed factor $e^{j\omega t}$ is ubiquitous and is suppressed. The media involved in the problem are linear and time-invariant and are considered to satisfy the following constitutive relations:

$$\begin{cases} \mathbf{D} = (1/c_0)\,P\,\mathbf{E} + L\,\mathbf{B} & \text{in } \Omega, \\ \mathbf{H} = M\,\mathbf{E} + c_0\,Q\,\mathbf{B} & \text{in } \Omega. \end{cases} \tag{1}$$

In the above equation, $\mathbf{E}$, $\mathbf{B}$, $\mathbf{D}$, and $\mathbf{H}$ are complex valued functions defined in Ω and represent, respectively, the electric field, magnetic induction, electric displacement, and magnetic field, while c_0 is the speed of light in a vacuum. The space where we will seek $\mathbf{E}$ and $\mathbf{H}$ is [12] (p. 82; see also p. 69):

$$U = H_{L^2,\Gamma}(\mathrm{curl}, \Omega) = \{\mathbf{v} \in H(\mathrm{curl}, \Omega) \mid \mathbf{v} \times \mathbf{n} \in L_t^2(\Gamma)\}, \tag{2}$$

where [12] (p. 48):

$$L_t^2(\Gamma) = \{\mathbf{v} \in (L^2(\Gamma))^3 \mid \mathbf{v} \cdot \mathbf{n} = 0 \text{ almost everywhere on } \Gamma\}. \tag{3}$$

Based on Maxwell's equations, boundary conditions, and constitutive relations, the following variational formulation of the problem can be deduced [5]: given $\omega > 0$, the electric and magnetic current densities prescribed by the sources $\mathbf{J}_e, \mathbf{J}_m \in (L^2(\Omega))^3$ and the known term $\mathbf{f}_R \in L_t^2(\Gamma)$, involved in admittance boundary condition, find $\mathbf{E} \in U$ such that:

$$a(\mathbf{E}, \mathbf{v}) = l(\mathbf{v}) \ \ \forall \mathbf{v} \in U, \tag{4}$$

where:

$$a(\mathbf{u}, \mathbf{v}) = c_0 \left(Q \operatorname{curl} \mathbf{u}, \operatorname{curl} \mathbf{v}\right)_{0,\Omega} - \frac{\omega^2}{c_0} \left(P \mathbf{u}, \mathbf{v}\right)_{0,\Omega} - j\omega \left(M \mathbf{u}, \operatorname{curl} \mathbf{v}\right)_{0,\Omega}$$

$$- j\omega \left(L \operatorname{curl} \mathbf{u}, \mathbf{v}\right)_{0,\Omega} + j\omega \left(Y \left(\mathbf{n} \times \mathbf{u} \times \mathbf{n}\right), \mathbf{n} \times \mathbf{v} \times \mathbf{n}\right)_{0,\Gamma}, \tag{5}$$

and:

$$l(\mathbf{v}) = -j\omega \left(\mathbf{J}_e, \mathbf{v}\right)_{0,\Omega} - c_0 \left(Q \mathbf{J}_m, \operatorname{curl} \mathbf{v}\right)_{0,\Omega} + j\omega \left(L \mathbf{J}_m, \mathbf{v}\right)_{0,\Omega} - j\omega \left(\mathbf{f}_R, \mathbf{n} \times \mathbf{v} \times \mathbf{n}\right)_{0,\Gamma}. \tag{6}$$

In [9], we derived a set of sufficient conditions that guarantee the well posedness and finite element approximability of the problem. The developed theory was applied to problems involving rotating axisymmetric objects. In this paper, we apply the theory to a wider range of problems involving bianisotropic materials and metamaterials demonstrating the generality of the developments and obtaining interesting new solutions [1,10,11].

We recall important definitions and hypotheses to fix the notations that are required for proceeding with the application of the theory. The subscript $i \in I$ identifying the subdomain Ω_i may belong to two subsets, I_a and I_b of I, according to the properties of the media involved: $i \in I_a$ when they are anisotropic (that is, with $L = M = 0$) and $i \in I_b$ when they are bianisotropic. Any matrix A with complex entries can be split into $A = A_s - jA_{ss}$ with $A_s = \frac{A+A^*}{2}$ and $A_{ss} = \frac{A^*-A}{2j}$. Some of the hypotheses will be stated using the alternative form of constitutive relations defined by:

$$\begin{cases} \mathbf{E} = \kappa \mathbf{D} + \chi \mathbf{B} & \text{in } \Omega, \\ \mathbf{H} = \gamma \mathbf{D} + \nu \mathbf{B} & \text{in } \Omega. \end{cases} \tag{7}$$

where the constitutive matrices are given by [13] $\kappa = c_0 P^{-1}$, $\chi = -c_0 P^{-1} L$, $\gamma = c_0 M P^{-1}$, and $\nu = c_0 \left(Q - M P^{-1} L\right)$.

As mentioned in Section 6 of [9], most of the hypotheses are readily satisfied for important practical problems. That leaves us with seven critical hypotheses (HM9–HM15 in [9]) on the media involved in the problem that need to be verified. Among them, HM9–HM12 of [9] are stated in terms of the constitutive relations involving κ, ν, χ, and γ and are restated here as H1–H4.

Hypothesis 1 (H1). $\exists \exists C_{\kappa,d} > 0, C_{\nu,d} > 0 : |determinant\,(\kappa)| \geq C_{\kappa,d}, |determinant\,(\nu)| \geq C_{\nu,d}, \forall \mathbf{x} \in \overline{\Omega}_i, \forall i \in I,$

Hypothesis 2 (H2). $\mathbf{1}_{1,3}^T \kappa^{-1} \mathbf{1}_{1,3} \neq 0,\ \mathbf{1}_{1,3}^T \nu^{-1} \mathbf{1}_{1,3} \neq 0\ \forall \mathbf{1}_{1,3} \in \mathbb{R}^3, \mathbf{1}_{1,3} \neq 0,\ \forall \mathbf{x} \in \overline{\Omega}_i,\ \forall i \in I_a,$

Hypothesis 3 (H3). $\exists \exists C_{\kappa,r} > 0,\ C_{\nu,r} > 0 : |\mathbf{1}_{1,3,n}^T \kappa^{-1} \mathbf{1}_{1,3,n}| \geq C_{\kappa,r},\ |\mathbf{1}_{1,3,n}^T \nu^{-1} \mathbf{1}_{1,3,n}| \geq C_{\nu,r}\ \forall \mathbf{1}_{1,3,n} \in \mathbb{R}^3 :$ $\|\mathbf{1}_{1,3,n}\|_2 = 1,\ \forall \mathbf{x} \in \overline{\Omega}_i,\ \forall i \in I_b,$

Hypothesis 4 (H4). $\exists \exists C_{\kappa,s} > 0, C_{\nu,s} > 0:$

$$\left(\sum_{i,j=1}^{3} |\kappa_{ij}| \right) - \min_{i=1,2,3} |\kappa_{ii}| \leq C_{\kappa,s} \quad \forall \mathbf{x} \in \overline{\Omega}_k,\ \forall k \in I_b, \tag{8}$$

$$\left(\sum_{i,j=1}^{3} |\nu_{ij}| \right) - \min_{i=1,2,3} |\nu_{ii}| \leq C_{\nu,s} \quad \forall \mathbf{x} \in \overline{\Omega}_k,\ \forall k \in I_b, \tag{9}$$

and κ, χ, γ, and ν satisfy:

$$\frac{4 \left(\left(\Sigma_{i,j=1}^{3} |\gamma_{ij}| \right) - \min_{i=1,2,3} |\gamma_{ii}| \right) \left(\left(\Sigma_{i,j=1}^{3} |\chi_{ij}| \right) - \min_{i=1,2,3} |\chi_{ii}| \right)}{\left(-C_{\kappa,s} + \sqrt{C_{\kappa,s}^2 + 4 C_{\kappa,d} C_{\kappa,r}} \right) \left(-C_{\nu,s} + \sqrt{C_{\nu,s}^2 + 4 C_{\nu,d} C_{\nu,r}} \right)} < 1 \tag{10}$$

$\forall \mathbf{x} \in \overline{\Omega}_k$, $\forall k \in I_b$.

Among the above hypotheses, H2 needs to hold only in the subdomains Ω_i, $i \in I_a$. H1 can be verified separately in any subdomain Ω_i, $i \in I$, whereas H3 and H4 are to be defined and verified only locally on any subdomain Ω_i, $i \in I_b$ (see Remark 1 of [9]).

The local continuity of the tensors P, Q, L, and M can be assumed in most practical problems, which allows the definition of the following constants.

- $\exists C_L > 0$: $|(L\,\mathrm{curl}\,\mathbf{u}, \mathbf{v})_{0,\Omega}| \leq C_L \|\mathrm{curl}\,\mathbf{u}\|_{0,\Omega} \|\mathbf{v}\|_{0,\Omega}$ for all $\mathbf{u} \in H(\mathrm{curl}, \Omega)$ and $\mathbf{v} \in (L^2(\Omega))^3$,
- $\exists C_M > 0$: $|(M\,\mathbf{u}, \mathrm{curl}\,\mathbf{v})_{0,\Omega}| \leq C_M \|\mathbf{u}\|_{0,\Omega} \|\mathrm{curl}\,\mathbf{v}\|_{0,\Omega}$ for all $\mathbf{u} \in (L^2(\Omega))^3$ and $\mathbf{v} \in H(\mathrm{curl}, \Omega)$.

Finally, HM13–HM15 of [9] are reported here as H5–H7.

Hypothesis 5 (H5). *We can find $C_{PS} > 0$ such that $|(P\mathbf{u}, \mathbf{u})_{0,\Omega}| \geq C_{PS}\|\mathbf{u}\|_{0,\Omega}^2$ for all $\mathbf{u} \in (L^2(\Omega))^3$.*

Hypothesis 6 (H6). *We can find $C_{QS} > 0$ such that $|(Q\mathrm{curl}\,\mathbf{u}, \mathrm{curl}\,\mathbf{u})_{0,\Omega}| \geq C_{QS}\|\mathrm{curl}\,\mathbf{u}\|_{0,\Omega}^2$ for all $\mathbf{u} \in H(\mathrm{curl}, \Omega)$.*

Hypothesis 7 (H7). *C_{PS}, C_{QS}, C_L, and C_M (i.e., all media involved) are such that $C_{QS} - \frac{C_L C_M}{C_{PS}} > 0$.*

We refer to Section 6 of [9] for some hints about the calculation of the constants involved in the above conditions. In particular, we recall here Lemma 1 of [9] for easy reference, which is helpful in finding the constant involved in H5.

Lemma 1. *Suppose that P_{ss} is uniformly positive definite in $\Omega_{el} \subset \Omega$, that is $\exists C_1 > 0$ such that:*

$$\int_{\Omega_{el}} \mathbf{u}^* P_{ss} \mathbf{u} \geq C_1 \int_{\Omega el} |\mathbf{u}|^2 = C_1 \|\mathbf{u}\|_{0,\Omega_{el}}^2 \ \forall \mathbf{u} \in (L^2(\Omega))^3. \tag{11}$$

whenever $\Omega_{el} = \Omega$, we can simply define $C_{PS} = C_1$.

Whenever Ω_{el} is not the whole Ω, suppose that, in the complementary region, P_s is uniformly positive or negative definite, that is $\exists C_5 > 0$ such that:

$$\left| \int_{\Omega \backslash \Omega_{el}} \mathbf{u}^* P_s \mathbf{u} \right| \geq C_5 \|\mathbf{u}\|_{0, \Omega \backslash \Omega_{el}}^2. \tag{12}$$

whenever $\Omega_{el} = \varnothing$, we simply have $C_{PS} = C_5$, and we can set:

$$C_{PS} = \min_{i \in I} \inf_{\mathbf{x} \in \Omega_i} \lambda_{min}(P_s), \tag{13}$$

where λ_{min} denotes the minimum of the magnitudes of the eigenvalues of the Hermitian symmetric matrix P_s. Finally, whenever Ω_{el} is neither the empty set nor the whole domain, under assumptions HM2 and HM3 of [9], condition H5 is satisfied with C_{PS} given by:

$$C_{PS} = \frac{1}{\sqrt{2}} \min\left(\sqrt{(1-\alpha)}C_5, \sqrt{C_1^2 + (1 - \frac{1}{\alpha})C_3^2} \right), \tag{14}$$

where $C_3 > 0$ is defined by:

$$\left| \int_{\Omega_{el}} \mathbf{u}^* P_s \mathbf{u} \right| \leq C_3 \|\mathbf{u}\|_{0,\Omega_{el}}^2 \tag{15}$$

and α is such that $1 > \alpha > \frac{C_3^2}{C_1^2 + C_3^2} > 0$.

Analogously, by replacing P with Q in Equations (11), (12), and (15), we define, respectively, Ω_{ml} and the constants $C_2 > 0$, $C_4 > 0$, and $C_6 > 0$ and deduce that condition H6 is satisfied if we set:

$$C_{QS} = \min_{i \in I} \ \inf_{\mathbf{x} \in \Omega_i} \lambda_{min}(Q_s),\tag{16}$$

whenever $\Omega_{ml} = \varnothing$, $C_{QS} = C_2$ whenever $\Omega_{ml} = \Omega$, or:

$$C_{QS} = \frac{1}{\sqrt{2}} \min \left(\sqrt{(1 - \alpha)C_6}, \sqrt{C_2^2 + (1 - \frac{1}{\alpha})C_4^2} \right),\tag{17}$$

α being such that $1 > \alpha > \frac{C_4^2}{C_2^2 + C_4^2} > 0$, when $\Omega_{ml} \neq \Omega$ and $\Omega_{ml} \neq \varnothing$.

With respect to the H4, if we define:

$$C_{\chi,s} = \max_{i \in I_b} \ \sup_{\mathbf{x} \in \Omega_i} \left(\left(\sum_{i,j=1}^{3} |\chi_{ij}| \right) - \min_{i=1,2,3} |\chi_{ii}| \right),\tag{18}$$

$$C_{\gamma,s} = \max_{i \in I_b} \ \sup_{\mathbf{x} \in \Omega_i} \left(\left(\sum_{i,j=1}^{3} |\gamma_{ij}| \right) - \min_{i=1,2,3} |\gamma_{ii}| \right),\tag{19}$$

the sufficient condition guaranteeing the validity of the inequality in the hypothesis can be expressed as:

$$K_u = \frac{4C_{\chi,s}C_{\gamma,s}}{\left(-C_{\kappa,s} + \sqrt{C_{\kappa,s}^2 + 4C_{\kappa,d}C_{\kappa,r}}\right)\left(-C_{\nu,s} + \sqrt{C_{\nu,s}^2 + 4C_{\nu,d}C_{\nu,r}}\right)} < 1.\tag{20}$$

3. Results and Discussion

In this section, we apply the theory developed in [9] to several classes of problems that could not be managed with the previous theories [5–8]. The conditions are established on the constitutive parameters of such problems, under which the well posedness and finite element approximability can be guaranteed. In particular, we apply the theory and obtain solutions for three examples of bianisotropic media, which are found in the open literature. The first one is that introduced in [1] where the authors considered plasmonic gratings, which are represented by an equivalent bianisotropic media. In this case, we study the scattering from a slab of the equivalent medium, which is placed in empty space, in accordance with the setup considered by the authors [1]. Next, the media introduced in [10,11] are analyzed. The authors there considered the scattering from the bianisotropic obstacles placed inside hollow waveguides. Although we stick to the original configurations proposed by the authors, our tools can be used to analyze other problems involving these media.

Under the conditions that guarantee that hypotheses H1–H7 are satisfied, the numerical solutions of these problems are computed. They can be used as reference solutions for other approaches and simulators because, on the one hand, for each problem, our theory guarantees the convergence of the sequence of approximations, and on the other hand, we verify that the outcome we present is close to the limit of the sequence by a stability analysis of the numerical solutions.

We use a first order edge element based Galerkin finite element simulator to obtain the solutions as described in Section 5 of [9].

3.1. Scattering from Plasmonic Gratings Behaving as Bianisotropic Metamaterials

In [1], the authors considered a plasmonic grating that exhibits bianisotropy at visible wavelengths. The bianisotropic media considered there are of the form:

$$\begin{cases} \mathbf{D} = \varepsilon\mathbf{E} + \xi\mathbf{H}, \\ \mathbf{B} = \zeta\mathbf{E} + \mu\mathbf{H}, \end{cases}\tag{21}$$

with:

$$\varepsilon = \varepsilon_0 \begin{bmatrix} \varepsilon_x & 0 & 0 \\ 0 & \varepsilon_y & 0 \\ 0 & 0 & \varepsilon_z \end{bmatrix}, \mu = \mu_0 \begin{bmatrix} \mu_x & 0 & 0 \\ 0 & \mu_y & 0 \\ 0 & 0 & \mu_z \end{bmatrix}, \tag{22}$$

$$\xi = \frac{1}{c_0} \begin{bmatrix} 0 & 0 & 0 \\ 0 & 0 & 0 \\ 0 & j\xi_0 & 0 \end{bmatrix}, \zeta = \frac{1}{c_0} \begin{bmatrix} 0 & 0 & 0 \\ 0 & 0 & -j\xi_0 \\ 0 & 0 & 0 \end{bmatrix}. \tag{23}$$

Here, ε_x, ε_y, and ε_z are complex valued functions and are set to be equal to a unique value ε_r. Moreover, μ_x, μ_y, and μ_z are each set equal to one, and ξ_0 is taken to be real valued. The region occupied by the scatterer may be denoted as $\Omega_s \subset \Omega$. The above form can be converted into the alternative form of constitutive relations [13] involving the P, Q, L, and M matrices defined in Equation (1), and the final result is shown in the following equations:

$$P = c_0 \varepsilon_0 \begin{bmatrix} \varepsilon_r & 0 & 0 \\ 0 & \varepsilon_r & 0 \\ 0 & 0 & \varepsilon_r - \xi_0^2 \end{bmatrix}, \tag{24}$$

$$Q = \frac{1}{c_0 \mu_0} I_3, \tag{25}$$

$$L = M^T = \frac{j\xi_0}{\mu_0 c_0} \begin{bmatrix} 0 & 0 & 0 \\ 0 & 0 & 0 \\ 0 & 1 & 0 \end{bmatrix}. \tag{26}$$

Here, I_3 is the three by three identity matrix. The complementary region $\Omega \setminus \Omega_s$ is occupied by the empty space, which is characterized by $P = c_0 \varepsilon_0 I_3$, $Q = \frac{1}{c_0 \mu_0} I_3$, $L = M = 0$. This problem cannot be managed by the previous theories and, in particular, by the theory in [5], which relied on strong hypothesis about losses.

The bianisotropic medium is lossy with the imaginary part $Im(\varepsilon_r) < 0$, whereas ξ_0 is assumed real here to avoid some longer calculations. Now, Lemma 1 can be applied to verify hypothesis H5. Inside Ω_s, P can be decomposed as $P = P_s - jP_{ss}$ with:

$$P_s = \frac{P + P^*}{2} = c_0 \varepsilon_0 \begin{bmatrix} Re(\varepsilon_r) & 0 & 0 \\ 0 & Re(\varepsilon_r) & 0 \\ 0 & 0 & Re(\varepsilon_r) - \xi_0^2 \end{bmatrix}, \tag{27}$$

and $P_{ss} = \frac{P^* - P}{2j} = -c_0 \varepsilon_0 Im(\varepsilon_r) I_3$. Hence, we have $\Omega_{el} = \Omega_s$, the lossy region where P_{ss} is uniformly positive definite and the complementary region with the free space where P_s is uniformly positive definite. This means that the conditions of Lemma 1 are satisfied, and as a result, H5 is valid.

From the definitions (see Equations (11), (12), and (15)), $C_1 = c_0 \varepsilon_0 |Im(\varepsilon_r)|$, $C_5 = c_0 \varepsilon_0$, and $C_3 = c_0 \varepsilon_0 \max(|Re(\varepsilon_r) - \xi_0^2|, |Re(\varepsilon_r)|)$. To find the minimum of the two expressions in Equation (14), we note that, in the valid range, the value of the first expression decreases monotonically with α, whereas that of the second expression increases with it. The highest estimate for C_{PS} is obtained when the two expressions have the same value. The value of α at which this happens can be evaluated by equating the two expressions and finding the positive root of the resulting quadratic equation. This value of α, denoted as α_{opt}, is given by:

$$\alpha_{opt} = \frac{C_5^2 - C_1^2 - C_3^2 + \sqrt{(C_5^2 - C_1^2 - C_3^2)^2 + 4C_5^2 C_3^2}}{2C_5^2}. \tag{28}$$

Thus, we may simply write:

$$C_{PS} = \sqrt{\frac{1 - \alpha_{opt}}{2}} c_0 \varepsilon_0. \tag{29}$$

As mentioned in [9], this does not mean that a better value of C_{PS} cannot be found. For example, if $Re(\varepsilon_r) - \xi_0^2 > 0$, then P_s is uniformly positive definite in Ω, and we can find another candidate for C_{PS}, namely $C_7 = c_0 \varepsilon_0 \min(1, |Re(\varepsilon_r) - \xi_0^2|)$ (see Equation (31) of [9]). In particular, when $Re(\varepsilon_r) - \xi_0^2 > \frac{1}{\sqrt{2}}$, C_7 is always going to give a value for C_{PS} that is higher than that obtained from the lemma.

In the rest of the subsection, the discussion focuses on the cases with $Re(\varepsilon_r) < 0$, which gives a non-definite P_s in Ω. For this case, $C_3 = c_0 \varepsilon_0 |Re(\varepsilon_r) - \xi_0^2|$, and we can directly use the value of C_{PS} in Equation (29). Since the material is assumed to be non-magnetic, the direct application of the definition gives $C_{QS} = \frac{1}{c_0 \mu_0}$, and H6 is valid. Likewise, for the continuity constants C_L and C_M, we can choose the value $\frac{|\xi_0|}{c_0 \mu_0}$. Then, the inequality in H7 becomes $C_{QS} - \frac{C_L C_M}{C_{PS}} = \frac{1}{c_0 \mu_0}\left(1 - \frac{\xi_0^2 \sqrt{2}}{\sqrt{1 - \alpha_{opt}}}\right) > 0$, which gives:

$$|\xi_0| < \left(\frac{1 - \alpha_{opt}}{2}\right)^{1/4}. \tag{30}$$

Since the right-hand side of Equation (30) also depends on ξ_0 due to the presence of C_3 in the expression for α_{opt}, we do not have a closed-form expression on the limit on ξ_0 below which H7 is satisfied. However, a graphical analysis can be done for estimating such a limit on $|\xi_0|$, as shown in Figure 1. The value of $Re(\varepsilon_r)$ is varied in the range $(-5.0, -1.0)$, whereas $Im(\varepsilon_r)$ assumes values in the range $(-0.5, -0.1)$. It can be observed that for a fixed value of $Im(\varepsilon_r)$, the range of ξ_0 over which H7 is valid steadily decreases as $|Re(\varepsilon_r)|$ increases. As for the dependence on $Im(\varepsilon_r)$, the corresponding range increases when the medium becomes lossier due to higher $|Im(\varepsilon_r)|$, as expected.

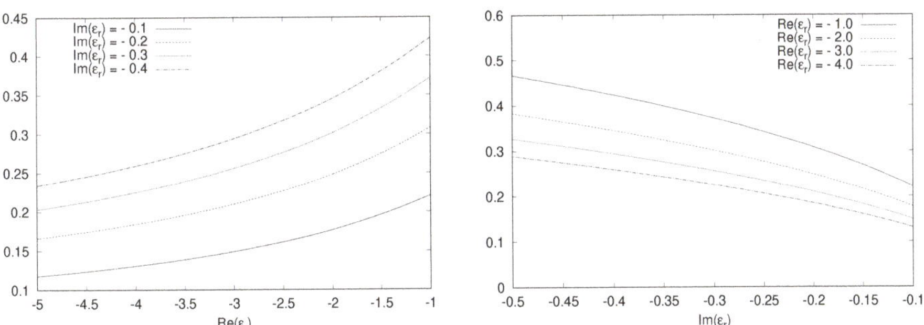

Figure 1. The plot indicates the maximum of $|\xi_0|$ guaranteeing that condition H7 is satisfied, for scattering problems involving different media considered in [1]. The curves are computed by assuming $\xi_0 \in \mathbb{R}$, $Re(\varepsilon_r) \in (-5.0, -1.0)$, $Im(\varepsilon_r) \in (-0.5, -0.1)$, and $\mu_r = 1$.

Since outside the region occupied by the bianisotropic media ($\Omega \setminus \Omega_s$), we just have the empty space, H1 and H2 are trivially satisfied there. By Remark 1 of [9], since H1, H3, and H4 need to hold only locally, now we have to just analyze them inside Ω_s occupied by the bianisotropic medium. We consider the alternative form of constitutive relations, which for the medium inside Ω_s becomes as in Equations (31) to (33), for examining the validity of H1, H3 and H4 [13]:

$$\kappa = \frac{1}{\varepsilon_0 \varepsilon_r} \begin{bmatrix} 1 & 0 & 0 \\ 0 & 1 & 0 \\ 0 & 0 & \frac{\varepsilon_r}{\varepsilon_r - \xi_0^2} \end{bmatrix}, \tag{31}$$

$$\nu = \frac{1}{\mu_0} \begin{bmatrix} 1 & 0 & 0 \\ 0 & \frac{\varepsilon_r}{\varepsilon_r - \zeta_0^2} & 0 \\ 0 & 0 & 1 \end{bmatrix}, \tag{32}$$

$$\gamma = -\chi^T = \frac{j\zeta_0 c_0}{\varepsilon_r - \zeta_0^2} \begin{bmatrix} 0 & 0 & 0 \\ 0 & 0 & 1 \\ 0 & 0 & 0 \end{bmatrix}. \tag{33}$$

The constants of interest can be evaluated directly from the definitions. The determinants of κ and ν are, respectively, $\dfrac{1}{(\varepsilon_0 \varepsilon_r)^3 (1 - \frac{\zeta_0^2}{\varepsilon_r})}$ and $\dfrac{1}{(\mu_0)^3 (1 - \frac{\zeta_0^2}{\varepsilon_r})}$, which immediately give the values of $C_{\kappa,d}$ and $C_{\nu,d}$.

$$C_{\kappa,d} = \frac{1}{|\varepsilon_0^3 \varepsilon_r^3 (1 - \frac{\zeta_0^2}{\varepsilon_r})|}, \tag{34}$$

$$C_{\nu,d} = \frac{1}{|\mu_0^3 (1 - \frac{\zeta_0^2}{\varepsilon_r})|}. \tag{35}$$

The inverses of the diagonal matrices κ and ν are just the diagonal matrices with the reciprocal entries. Applying Equations (40) and (41) of [9] gives the values of $C_{\kappa,r}$ and $C_{\nu,r}$.

$$C_{\kappa,r} = |\varepsilon_0 \varepsilon_r|, \tag{36}$$

$$C_{\nu,r} = \mu_0. \tag{37}$$

Using Equations (36) and (37) of [9], we get $C_{\kappa,s}$ and $C_{\nu,s}$.

$$C_{\kappa,s} = \frac{2}{|\varepsilon_0 \varepsilon_r|}, \tag{38}$$

$$C_{\nu,s} = \frac{2}{\mu_0}. \tag{39}$$

From Equations (18) and (19), we can easily evaluate $C_{\gamma,s}$ and $C_{\chi,s}$.

$$C_{\chi,s} = C_{\gamma,s} = \left| \frac{\zeta_0 c_0}{\varepsilon_r - \zeta_0^2} \right|. \tag{40}$$

The hypotheses H1 and H3 are valid due to the existence of the above constants. Using these constants, the value of K_u can be calculated from Equation (20). The critical value of $|\zeta_0|$ below which the condition in H4 is satisfied is plotted in Figure 2, with respect to either $Re(\varepsilon_r)$ or $Im(\varepsilon_r)$. The results show that the range of ζ_0 for which H4 holds true increases with the increase in $|Re(\varepsilon_r)|$, while it is practically independent of $Im(\varepsilon_r)$.

Let us try to understand the implications of the theory by applying it to the numerical solution of a specific problem involving the medium of interest. We consider the region with the scatterer Ω_s to be a cube filled with homogeneous bianisotropic media. The surrounding region is filled with empty space, and the overall domain of numerical investigation, Ω, has a cubic shape as well and is concentric to Ω_s. In the following, Ω and Ω_s are characterized by sides of length 2 μm and 0.8 μm, respectively. The axes are taken along the sides of the cubic domain Ω, and the excitation is with a plane wave incident along the x axis, with the electric field polarized along the z axis, having a magnitude of 1 V/m and wavelength of 1 μm.

Inside Ω_s, the medium is characterized by $\varepsilon_r = -1 - j0.4$, $\mu_r = 1$, and $\zeta_0 = -0.41$. This value is such that the hypotheses required for well posedness and finite element approximability are satisfied.

In fact, for the ε_r considered, condition H4 is valid for $|\xi_0| < 0.4393$, and condition H7 is valid for $|\xi_0| < 0.4235$.

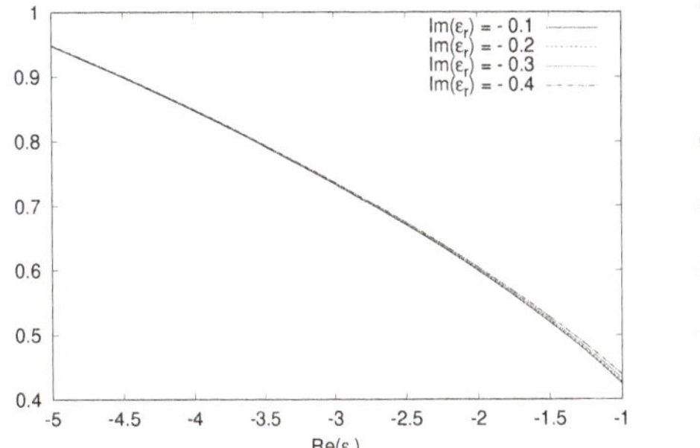
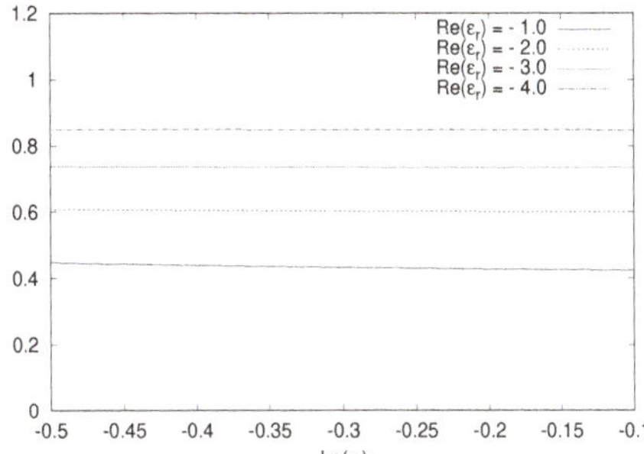

Figure 2. The plots indicate the maximum $|\xi_0|$ guaranteeing that condition H4 is satisfied, for scattering problems involving different media considered in [1]. The curves are computed by assuming $\xi_0 \in \mathbb{R}$, $Re(\varepsilon_r) \in (-5.0, -1.0)$, $Im(\varepsilon_r) \in (-0.5, -0.1)$, and $\mu_r = 1$.

The solutions are obtained with a first order edge element based Galerkin finite element method. The boundary condition is enforced with Y equal to the admittance of a vacuum and with an inhomogeneous term $\mathbf{f}_R$, taking into account the incident field.

The domain is discretized uniformly using tetrahedral meshes. The meshing is done by first dividing the domain into small identical cubes, each of which is in turn divided into six tetrahedra. The parameter h denotes the maximum diameter of all the elements of the mesh [14] (p. 131), and in this case, it is simply given by the side of the small cubes times $\sqrt{3}$. To study the stability of the solution, we consider different levels of refinement of meshes ranked in order of h, ranging from "very coarse" to "very fine". For example, the mesh denoted as very coarse is characterized by cubes of sides 200 nm, and the resulting mesh has 1331 nodes, 6000 tetrahedral elements, and 1200 boundary faces. A summary of the information related to the four different refinements of the meshes that were used is given in Table 1.

Table 1. Details of the different meshes used.

Type of Mesh	Maximum Diameter of the Mesh (h in nm)	Number of Nodes	Number of Elements	Number of Boundary Faces
Very coarse	$200\ \sqrt{3}$	1331	6000	1200
Coarse	$100\ \sqrt{3}$	9261	48,000	4800
Fine	$50\ \sqrt{3}$	68,921	384,000	19,200
Very fine	$25\ \sqrt{3}$	531,441	3,072,000	76,800

The outcomes related to the stability of the results of the simulations are shown in Figure 3 by plotting the magnitude of the z component of the electric field along a line parallel to the y axis and passing through the center of gravity of the domain. The difference between successive refinements progressively decreases, and the fine and very fine meshes give stable solutions. The well posedness and convergence result that was predicted using the theory guarantee that our solutions are reliable.

Figure 4 shows the significance of the bianisotropic effect on the z component of the electric field along a line parallel to the x axis and passing through the center of gravity of the domain. Here, E_z denotes the solution obtained with $\xi_0 = -0.41$, and $E_{z,0}$ is the solution when $\xi_0 = 0$, while the plot shows the magnitude of the difference $|E_z - E_{z,0}|$ along with $|E_z|$. The magnitude of the difference is as large as 50% of the incident field. Similarly, the results along a line parallel to the

z axis and passing through the center of gravity of the domain are shown in Figure 5, and we get a difference of around 30% of the incident field.

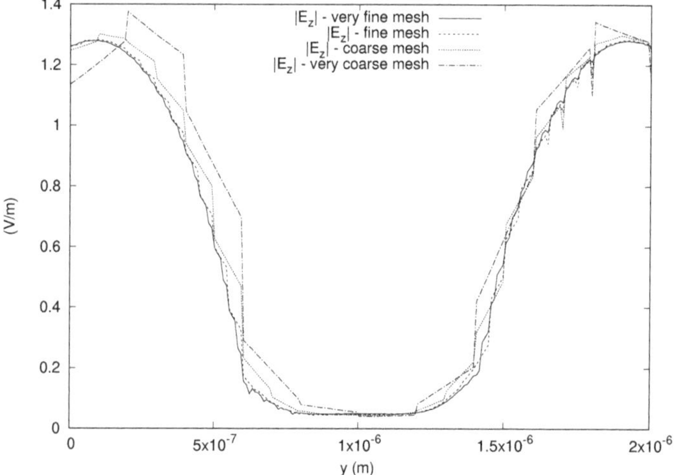

Figure 3. Stability of the solution for problem involving the medium in [1]. The magnitude of the z component of the electric field is plotted for four different meshes along a line parallel to the y axis and passing through the center of gravity of the domain.

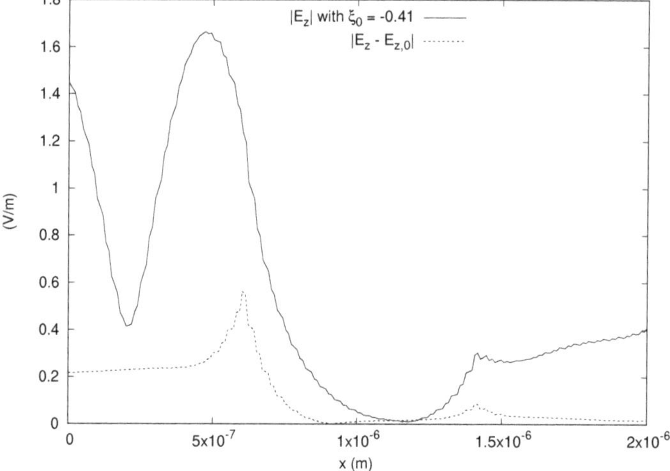

Figure 4. The magnitude of the z component of electric field along a line parallel to the x axis and passing through the center of gravity of the domain, for the problem involving the medium in [1]. The plot for the magnitude of the field $|E_z|$ obtained in the bianisotropic case using $\zeta_0 = -0.41$ is shown along with the magnitude of the difference between the two solutions $|E_z - E_{z,0}|$, where $E_{z,0}$ is obtained using $\zeta_0 = 0$.

These non-negligible effects imply that to get accurate results, it is necessary to consider the bianisotropy of the medium. Hence, the reliability of the finite element solution in the presence of bianisotropy is important for getting good results for these problems. The application of our theory gives the conditions under which we can guarantee such reliability. The solutions obtained can serve as references for other approaches.

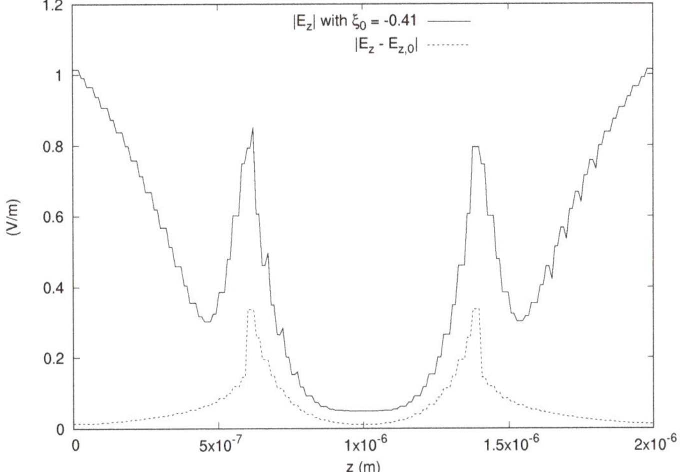

Figure 5. The magnitude of the z component of electric field along a line parallel to the z axis and passing through the center of gravity of the domain, for the problem involving the medium in [1]. The plot for the magnitude of the field $|E_z|$ obtained in the bianisotropic case using $\xi_0 = -0.41$ is shown along with the magnitude of the difference between the two solutions $|E_z - E_{z,0}|$, where $E_{z,0}$ is obtained using $\xi_0 = 0$.

3.2. Scattering from Chiral Obstacles in a Waveguide

In [10], the authors considered a metallic waveguide, which was hollow except for an obstacle characterized by a chiral medium with the following constitutive relations.

$$\begin{cases} \mathbf{D} = \varepsilon_0 \varepsilon_r I_3 \mathbf{E} - j\xi_c I_3 \mathbf{B}, \\ \mathbf{H} = -j\xi_c I_3 \mathbf{E} + \frac{1}{\mu_0 \mu_r} I_3 \mathbf{B}. \end{cases} \tag{41}$$

Here, ε_r, μ_r, and ξ_c are strictly positive real quantities. Thus, from Equation (1), we can easily identify P, Q, L, and M, which are given below:

$$P = \varepsilon_0 \varepsilon_r c_0 I_3, \tag{42}$$

$$Q = \frac{1}{\mu_0 \mu_r c_0} I_3, \tag{43}$$

$$L = M = -j\xi_c I_3. \tag{44}$$

In Section 6 of [5], it was shown that this media could not be managed by the theory developed there, independently of the value of $\xi_c \in \mathbb{R}$ and of any other material involved in the model of interest. However, we show that the generality of the recently developed theory in [9] allows us to apply it to obtain the conditions for well posedness and finite element approximability for this kind of problem of practical interest.

Let us analyze the validity of the hypotheses by considering, as did the authors in [10], $\varepsilon_r \geq 1$ and $\mu_r = 1$. We can make use of Lemma 1 of [9] to check H5. $P_s = P$ and is equal to $\varepsilon_0 \varepsilon_r c_0$ inside the material and simply $\varepsilon_0 c_0$ outside. Since $\Omega_{el} = \varnothing$, a value of C_{PS} can be found by using Equation (13). In particular, H5 is satisfied for $C_{PS} = \varepsilon_0 c_0$. The hypothesis H6 is also trivially valid with $C_{QS} = \frac{1}{\mu_0 c_0}$.

By Equations (32) and (33) of [9], $C_L = C_M = \xi_c$. Then, the inequality in hypothesis H7 becomes $C_{QS} - \frac{C_L C_M}{C_{PS}} = c_0(\varepsilon_0 - \mu_0 \xi_c^2) > 0$, which implies:

$$\xi_c < \sqrt{\frac{\varepsilon_0}{\mu_0}} = 2.654 \times 10^{-3} \text{ mho}. \tag{45}$$

This is not a small value considering the chiral effects reported in [10]. As the region outside the obstacle is empty space, H2 is trivially satisfied, and by Remark 1 of [9], we need to verify that the hypotheses H1, H3, and H4 hold true locally inside the region occupied by the bianisotropic medium. To do this, the suitable form of constitutive relations is in terms of κ, ν, γ, and χ, which are given by the following [13]:

$$\kappa = \frac{1}{\varepsilon_0 \varepsilon_r} I_3, \tag{46}$$

$$\nu = \frac{\varepsilon_0 \varepsilon_r + \mu_0 \xi_c^2}{\mu_0 \varepsilon_0 \varepsilon_r} I_3, \tag{47}$$

$$\chi = -\gamma = \frac{j \xi_c}{\varepsilon_0 \varepsilon_r} I_3. \tag{48}$$

κ and ν are multiples of the identity matrix with eigenvalues $\frac{1}{\varepsilon_0 \varepsilon_r}$ and $\left(\frac{\varepsilon_0 \varepsilon_r + \mu_0 \xi_c^2}{\mu_0 \varepsilon_0 \varepsilon_r}\right)$, respectively. The determinants are just the cubes of the eigenvalues, and hence, according to Equations (34) and (35) of [9], we get the values of $C_{\kappa,d}$ and $C_{\nu,d}$.

$$C_{\kappa,d} = \left(\frac{1}{\varepsilon_0 \varepsilon_r}\right)^3, \tag{49}$$

$$C_{\nu,d} = \left(\frac{\varepsilon_0 \varepsilon_r + \mu_0 \xi_c^2}{\mu_0 \varepsilon_0 \varepsilon_r}\right)^3. \tag{50}$$

$C_{\kappa,s}$ and $C_{\nu,s}$, by Equations (36) and (37) of [9], are in this case simply twice the eigenvalue of the corresponding diagonal matrix:

$$C_{\kappa,s} = \frac{2}{\varepsilon_0 \varepsilon_r}, \tag{51}$$

$$C_{\nu,s} = 2\frac{\varepsilon_0 \varepsilon_r + \mu_0 \xi_c^2}{\mu_0 \varepsilon_0 \varepsilon_r}. \tag{52}$$

The inverse of the matrices is also trivial, and Equations (40) and (41) of [9] simply evaluate to the reciprocals of the eigenvalues of κ and ν, respectively giving $C_{\kappa,r}$ and $C_{\nu,r}$:

$$C_{\kappa,r} = \varepsilon_0 \varepsilon_r, \tag{53}$$

$$C_{\nu,r} = \frac{\mu_0 \varepsilon_0 \varepsilon_r}{\varepsilon_0 \varepsilon_r + \mu_0 \xi_c^2}. \tag{54}$$

From Equations (18) and (19), we get:

$$C_{\chi,s} = C_{\gamma,s} = \frac{2\xi_c}{\varepsilon_0 \varepsilon_r}. \tag{55}$$

Having shown that the hypotheses H1 and H3 are satisfied, we can use the above constants to calculate K_u to verify H4. Figure 6 shows the dependence of K_u on ξ_c for various values of ε_r. As the value of ε_r increases, the hypothesis H4 remains valid for higher and higher values of ξ_c. Figure 7 shows the plot of the critical value of ξ_c below which H4 is satisfied against ε_r. The limit of 2.654×10^{-3} mho, arising from Equation (45) required to satisfy H7, is also shown in the same figure. It is seen that for low values of ε_r, the tighter condition arises from the need to satisfy H4. For example, the limiting value is 5.6×10^{-4} mho for $\varepsilon_r = 1$, increases with ε_r, and is 1.78×10^{-3} mho for $\varepsilon_r = 10$. The curve crosses the 2.654×10^{-3} mho line at around $\varepsilon_r = 22.3$, and above that value, Equation (45) imposes the stricter limit.

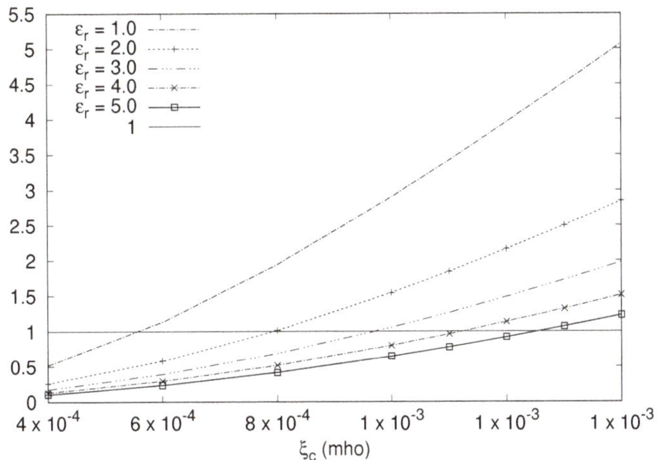

Figure 6. Plot of K_u versus ξ_c for the bianisotropic medium described in [10]. The plots are shown for various values of ε_r. The hypothesis H4 is satisfied for $K_u < 1$.

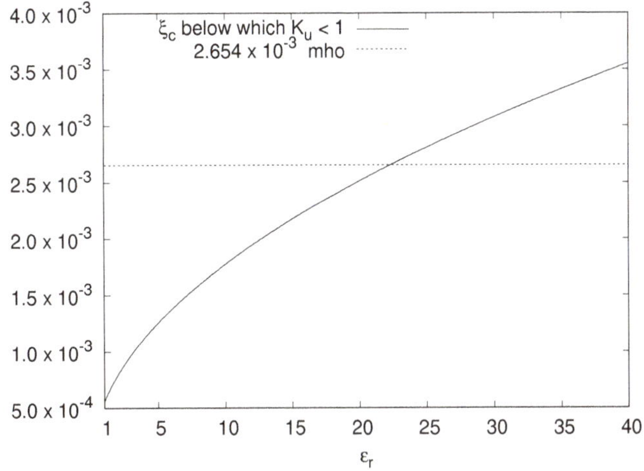

Figure 7. The value of ξ_c below which the hypothesis H4 is satisfied is plotted against ε_r. The limit of 2.654×10^{-3} mho, arising from Equation (45) required to satisfy H7, is also shown.

Now, we consider a specific numerical problem for which the solution is calculated using our finite element simulator. A rectangular waveguide with a discontinuity due to a block of bianisotropic medium is considered as shown in Figure 8. In the simulation, the rectangular waveguide is characterized by $a = 23$ mm, $b = 10$ mm and has a length $l = 40$ mm. The obstacle is a parallelepiped with $c = 11$ mm, $d = 5$ mm and a length $w = 10$ mm. The origin of the axis is at the lower right corner of the near face of the waveguide shown in Figure 8. The obstacle ranges from $x = 6$ mm to $x = 17$ mm along the x axis, from $y = 0$ to $y = 5$ mm along the y axis, and from $z = 15$ mm to $z = 25$ mm along the z axis. The bianisotropic medium making up the obstacle is characterized by $\varepsilon_r = 5$ and $\xi_c = 1.24 \times 10^{-3}$ mho. For this medium, $K_u = 0.98 < 1$, and also, Equation (45) is satisfied; hence, all the hypotheses required to guarantee the well posedness and convergence of finite element solutions hold true. The waveguide is excited with TE_{10} mode having an amplitude of 1 V/m and a frequency of 9 GHz. The input port is on the x-y plane, and the output port is closed on a matched homogeneous admittance boundary condition for the TE_{10} mode in the empty waveguide.

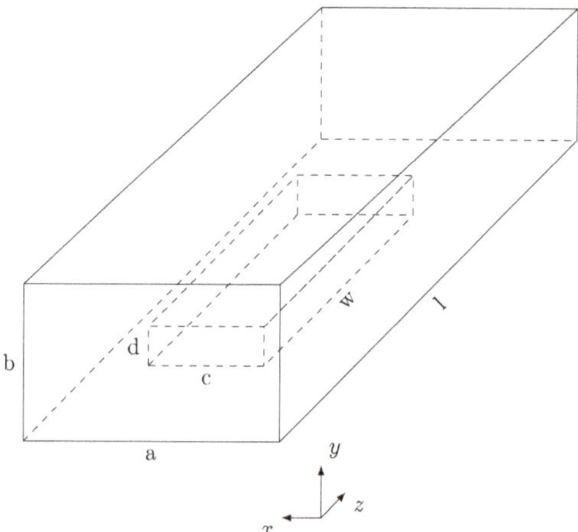

Figure 8. The geometry of a rectangular waveguide partially filled with the chiral media considered in [10].

The details of the Galerkin finite element solver is the same as before. The tetrahedral meshes are obtained as discussed in the previous subsection by dividing the domain into small cubes, each of which is in turn subdivided into six tetrahedra. The stability of the solution is verified by checking the solutions for three different meshes, which are characterized by small cubes of sides $\frac{1}{2}$ mm, $\frac{1}{4}$ mm, and $\frac{1}{6}$ mm, which are referred to as, respectively, "coarse", "fine", and "very fine" meshes. There are 10,824 nodes, 55,200 elements, and 6200 boundary faces in the coarse mesh, whereas the fine mesh has 79,947 nodes, 441,600 elements, and 24,800 boundary faces, and finally, the very fine mesh has 262,570 nodes, 1,490,400 elements, and 55,800 boundary faces. It was verified that the solutions obtained with these meshes are stable. For example, Figure 9 shows the magnitude of the x component of the electric field along a line parallel to the y axis and passing through the center of gravity of the domain with the different meshes and illustrates the stability of the result. It is noted that the x component of the electric field along this line is zero for the achiral case ($\xi_c = 0$), and there is a difference of more than 30% of the incident field, which is induced by the bianisotropy.

Figure 10 shows the result for the magnitude and phase of the x component of the electric field along the line parallel to the x axis and passing through the center of gravity of the domain. We have a difference of 20% of the magnitude of the incident field for the x component of the electric field along this line. Similarly, Figure 11 shows that the bianisotropic effect for the x component of the field along the line parallel to the z axis and passing through the center of gravity of the domain causes a difference in the magnitude of the electric field of around 13% of the incident field.

Figure 12 shows the y component of the electric field along a line parallel to the z axis and passing through the center of gravity of the domain. The bianisotropic effect is again not negligible and causes a difference of more than 10% of the incident field. A similar effect is present in the z component as can be seen in the plot along a line parallel to the x axis and passing through the center of gravity of the domain, which is given in Figure 13. We do not show the other figures to save space, but it is noted that the y component of electric field along the lines through the center of the domain and parallel to the x and y axes shows small differences in magnitude between the chiral and achiral cases, being less than five percent of the incident field. The z component on the other hand along the lines parallel to the y and z axes and passing through the center of the domain shows a difference of more than 15% of the incident field.

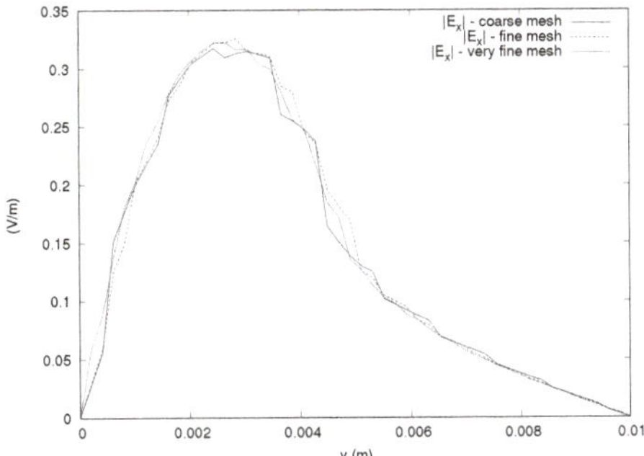

Figure 9. Stability of the solution for the problem involving the medium in [10]. The magnitude of the x component of the electric field is plotted for three different meshes along a line parallel to the y axis and passing through the center of gravity of the domain.

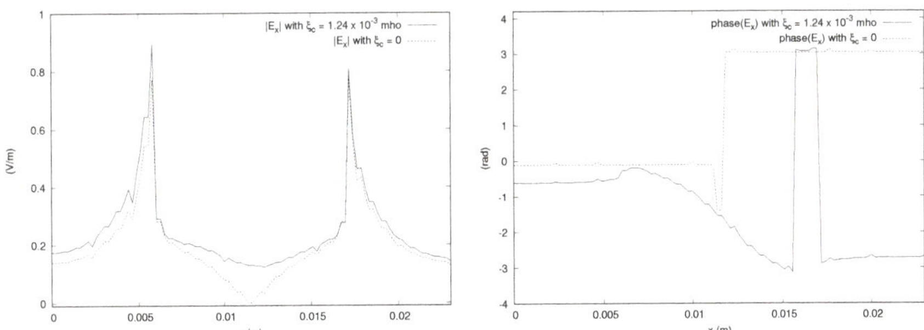

Figure 10. The magnitude and phase of the x component of the electric field along a line parallel to the x axis and passing though the center of gravity of the domain for the problem involving the medium in [10]. The plot for the bianisotropic case using $\xi_c = 1.24 \times 10^{-3}$ mho is compared with the solution obtained in the isotropic case using $\xi_c = 0$.

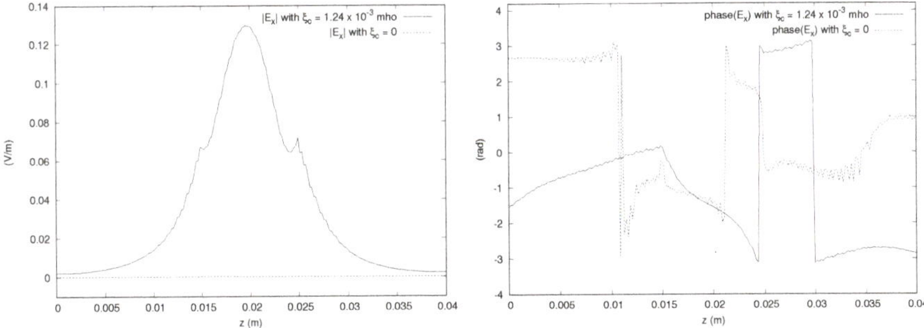

Figure 11. The magnitude and phase of the x component of the electric field along a line parallel to the z axis and passing though the center of gravity of the domain for the problem involving the medium in [10]. The plot for the bianisotropic case using $\xi_c = 1.24 \times 10^{-3}$ mho is compared with the solution obtained in the isotropic case using $\xi_c = 0$.

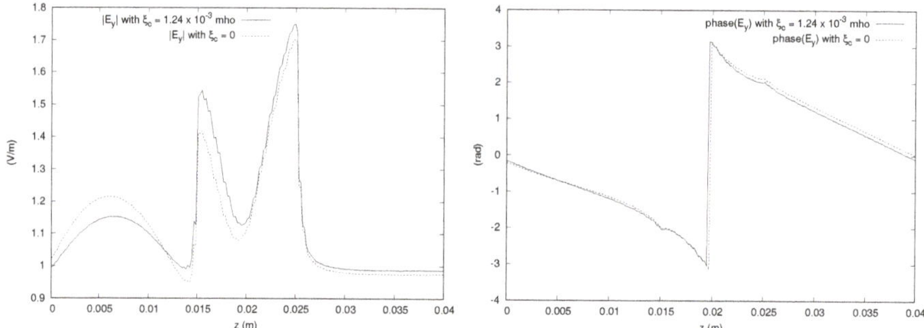

Figure 12. The magnitude and phase of the y component of the electric field along a line parallel to the z axis and passing though the center of gravity of the domain for the problem involving the medium in [10]. The plot for the bianisotropic case using $\xi_c = 1.24 \times 10^{-3}$ mho is compared with the solution obtained in the isotropic case using $\xi_c = 0$.

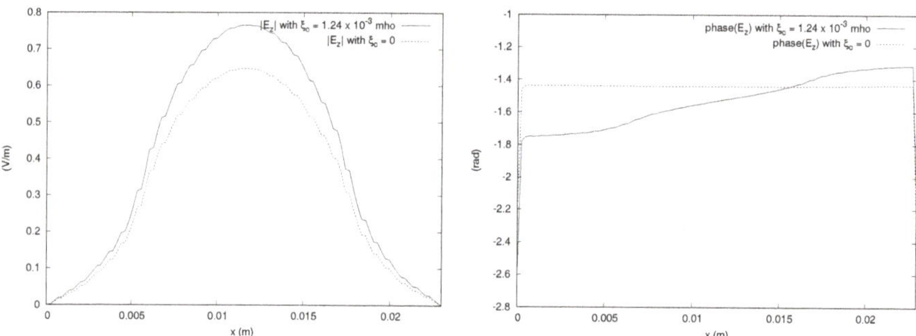

Figure 13. The magnitude and phase of the z component of the electric field along a line parallel to the x axis and passing though the center of gravity of the domain for problem involving the medium in [10]. The plot for the bianisotropic case using $\xi_c = 1.24 \times 10^{-3}$ mho is compared with the solution obtained in the isotropic case using $\xi_c = 0$.

Together, these results provide a point of reference for other approaches of solving such problems, owing to the reliability of the results provided here, which is guaranteed by the recently developed theory. The previous theory [5] was not able to manage these problems, and our results are therefore novel. In fact, to the best of our knowledge, there are no other approaches that are available to get reliable results for benchmarking. The significant bianisotropic effects demonstrated in the results show the practical importance of the theory for such media.

3.3. Reflection by a Short-Circuited Waveguide Half Filled with Bianisotropic Media

Another relevant bianisotropic medium was discussed in [11], where the authors considered a short-circuited rectangular waveguide, half of which was empty and the other half filled with a lossless bianisotropic material characterized by:

$$P = \varepsilon_0 c_0 I_3, \tag{56}$$

$$Q = \frac{1}{\mu_0 c_0} I_3, \tag{57}$$

$$L = M = j\kappa_0 A, \tag{58}$$

where A is the matrix given by:

$$A = \begin{bmatrix} 1 & 1 & 0 \\ 1 & 1 & 0 \\ 0 & 0 & 1 \end{bmatrix}, \tag{59}$$

and κ_0 is a positive real number.

The hypotheses H5 and H6 are trivially valid with $C_{PS} = \varepsilon_0 c_0$ and $C_{QS} = \frac{1}{\mu_0 c_0}$. Further, $L^*L = M^*M = \kappa_0^2 A^2$ whose eigenvalues are zero, κ_0^2, and $4\kappa_0^2$. Therefore, by Equations (32) and (33) of [9], we get $C_L = C_M = 2\kappa_0$. The condition in hypothesis H7 then becomes $C_{QS} - \frac{C_L C_M}{C_{PS}} = \frac{1}{\mu_0 c_0} - \frac{4\kappa_0^2}{\varepsilon_0 c_0} > 0$, which gives the following limit on κ_0:

$$\kappa_0 < \frac{1}{2}\sqrt{\frac{\varepsilon_0}{\mu_0}} = 1.327 \times 10^{-3} \text{ mho} . \tag{60}$$

The hypothesis H2 holds true since the anisotropic part is just empty space. H1, H3, and H4 can be studied using the alternative form of constitutive relations, which is characterized by the following matrices [13]:

$$\kappa = \frac{1}{\varepsilon_0} I_3, \tag{61}$$

$$\nu = \frac{1}{\mu_0} I_3 + \frac{\kappa_0^2}{\varepsilon_0} A^2, \tag{62}$$

$$\chi = -\gamma = \frac{-j\kappa_0}{\varepsilon_0} A. \tag{63}$$

The determinants of κ and ν can be readily calculated, and by using Equations (34) and (35) of [9], we get $C_{\kappa,d}$ and $C_{\nu,d}$:

$$C_{\kappa,d} = \frac{1}{\varepsilon_0^3}, \tag{64}$$

$$C_{\nu,d} = \frac{(\varepsilon_0 + \mu_0 \kappa_0^2)(\varepsilon_0 + 4\mu_0 \kappa_0^2)}{\mu_0^3 \varepsilon_0^2}. \tag{65}$$

$C_{\kappa,s}$ and $C_{\nu,s}$ can be directly obtained from their definitions:

$$C_{\kappa,s} = \frac{2}{\varepsilon_0}, \tag{66}$$

$$C_{\nu,s} = \frac{2\varepsilon_0 + 8\mu_0 \kappa_0^2}{\mu_0 \varepsilon_0}. \tag{67}$$

By simple application of the definition:

$$C_{\kappa,r} = \varepsilon_0 \tag{68}$$

and by Equation (41) of [9], $C_{\nu,r}$ evaluates to the reciprocal of the largest eigenvalue of the real matrix ν, which is given by:

$$C_{\nu,r} = \frac{\mu_0 \varepsilon_0}{\varepsilon_0 + 4\mu_0 \kappa_0^2}. \tag{69}$$

Equations (18) and (19) give:

$$C_{\chi,s} = C_{\gamma,s} = \frac{4\kappa_0}{\varepsilon_0}. \tag{70}$$

The existence of the above constants verifies H1 and H3, and we can use them to calculate K_u to check H4. It can be verified that K_u is less than one when $\kappa_0 \leq 2.72 \times 10^{-4}$ mho, which is stricter than the limit obtained from Equation (60).

Finally, let us give some numerical solutions for this problem, which can be used as references for other approaches. The cross-section of the waveguide is 2 cm along the x axis and 1 cm along the y axis, and the length of the waveguide is 2 cm, half of which is filled with the bianisotropic medium characterized by $\kappa_0 = 2.7 \times 10^{-3}$ mho (see Figure 1 of [11]). The origin is taken on the corner of the open face of the waveguide on the empty side. TE_{10} mode is excited in the waveguide with a source of amplitude 1 V/m and a frequency of 12 GHz.

The first order edge element based Galerkin finite element method is used to obtain the solution. The meshing is carried out by dividing the domain into identical cubes, each of which is then subdivided into six tetrahedra. The stability of the result is ensured by evaluating the solutions on three meshes, termed as "coarse", "fine", and "very fine", characterized, respectively, by cubes of sides 1 mm, $\frac{1}{2}$ mm, and $\frac{1}{3}$ mm. The coarse mesh has 4852 nodes, 24,000 elements, and 3200 boundary faces. The fine mesh is composed of 35,301 nodes, 192,000 elements, and 12,800 boundary faces. The very fine mesh has 115,351 nodes, 648,000 elements, and 28,800 boundary faces. The results obtained with these meshes are stable. For example, Figure 14 shows the stability of the results obtained with the three meshes for the x component of the electric field along a line parallel to the y axis and passing through the center of gravity of the domain.

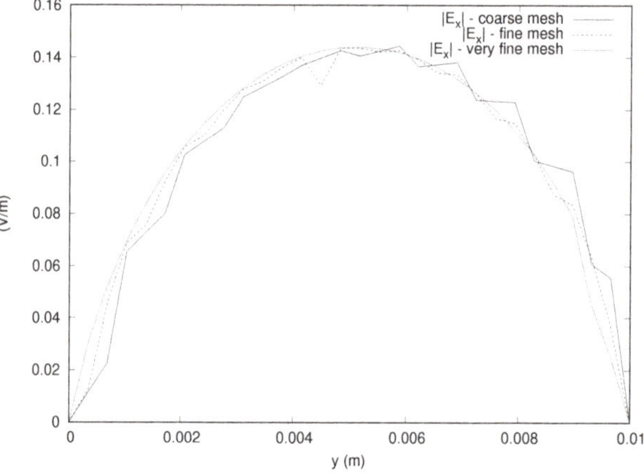

Figure 14. Stability of the solution for the problem involving the medium in [11]. The magnitude of the *x* component of the electric field is plotted for three different meshes along a line parallel to the *y* axis and passing through the center of gravity of the domain.

We provide the magnitudes and phases of the x component of the electric field obtained from the simulation in Figures 15–17. The bianisotropy causes a difference in magnitude of up to 14% of the incident field. The phases are also significantly affected by the bianisotropy of the medium. The figures for other components are not shown to save space, but it is noted that the y component is affected by the bianisotropy, showing a difference of up to about 10% of the incident field. The z component of the field along the line parallel to the *x* axis passing through the center of gravity of the domain does show a difference of about 10% of the incident field, but the magnitudes along the other directions are close to zero for both values of κ_0 considered.

Since the theory guarantees the reliability of these results, they can be used as references for other solvers and approaches. It is to be noted that the previous theory developed in [5] could not be applied to this medium since the same reasons mentioned there with respect to the medium in [10] are valid. Hence, our results show the generality of the new theory with respect to the application to interesting bianisotropic media. The results demonstrate that our theory can be applied to problems with significant bianisotropic effects. This consideration underlines its practical importance.

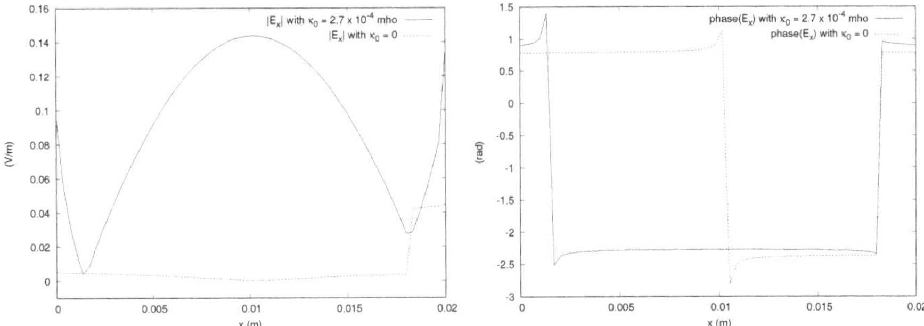

Figure 15. The magnitude and phase of the x component of the electric field along a line parallel to the x axis and passing though the center of gravity of the domain for the problem involving the medium in [11]. The plot for the bianisotropic case using $\kappa_0 = 2.7 \times 10^{-4}$ mho is compared with the solution obtained in the isotropic case using $\kappa_0 = 0$.

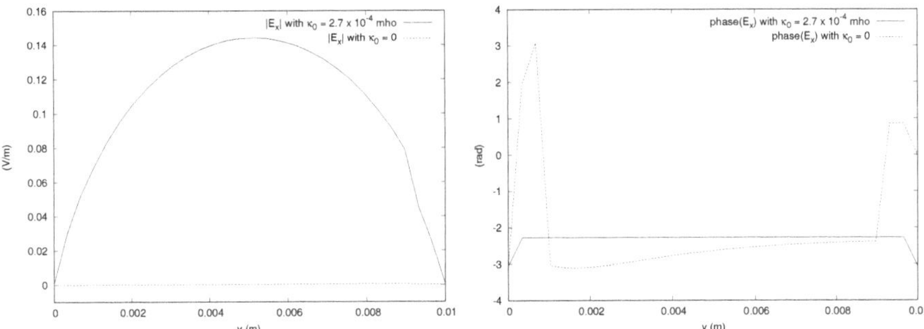

Figure 16. The magnitude and phase of the x component of the electric field along a line parallel to the y axis and passing though the center of gravity of the domain for the problem involving the medium in [11]. The plot for the bianisotropic case using $\kappa_0 = 2.7 \times 10^{-4}$ mho is compared with the solution obtained in the isotropic case using $\kappa_0 = 0$.

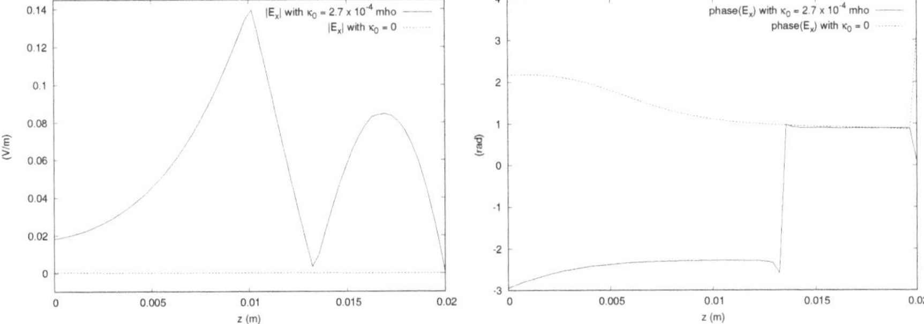

Figure 17. The magnitude and phase of the x component of the electric field along a line parallel to the z axis and passing though the center of gravity of the domain for the problem involving the medium in [11]. The plot for the bianisotropic case using $\kappa_0 = 2.7 \times 10^{-4}$ mho is compared with the solution obtained in the isotropic case using $\kappa_0 = 0$.

4. Conclusions

In this paper, we discussed the application of a recently developed theory to electromagnetic scattering problems involving bianisotropic materials and metamaterials of practical interest. The range of constitutive parameters over which the reliability of the results was guaranteed was calculated for these problems, and the solutions were obtained. The ensured reliability of the results made them useful as benchmarks for other numerical techniques. To the best of our knowledge, none of the previous tools were able to get such benchmark solutions.

Author Contributions: Both authors have contributed equally to this paper. All authors have read and agreed to the published version of the manuscript.

Funding: This research received no external funding.

Conflicts of Interest: The authors declare no conflict of interest.

References

1. Kraft, M.; Braun, A.; Luo, Y.; Maier, S.A.; Pendry, J.B. Bianisotropy and magnetism in plasmonic gratings. *ACS Photonics* **2016**, *3*, 764–769. [CrossRef]
2. Yazdi, M.; Albooyeh, M.; Alaee, R.; Asadchy, V.; Komjani, N.; Rockstuhl, C.; Simovski, C.R.; Tretyakov, S. A bianisotropic metasurface with resonant asymmetric absorption. *IEEE Trans. Antennas Propag.* **2015**, *63*, 3004–3015. [CrossRef]
3. Kildishev, A.V.; Borneman, J.D.; Ni, X.; Shalaev, V.M.; Drachev, V.P. Bianisotropic effective parameters of optical metamagnetics and negative-index materials. *Proc. IEEE* **2011**, *99*, 1691–1700. [CrossRef]
4. Kriegler, C.E.; Rill, M.S.; Linden, S.; Wegener, M. Bianisotropic photonic metamaterials. *IEEE J. Sel. Top. Quantum Electron.* **2010**, *16*, 367–375. [CrossRef]
5. Fernandes, P.; Raffetto, M. Well-posedness and finite element approximability of time-harmonic electromagnetic boundary value problems involving bianisotropic materials and metamaterials. *Math. Model. Methods Appl. Sci.* **2009**, *19*, 2299–2335. [CrossRef]
6. Brignone, M.; Raffetto, M. Well posedness and finite element approximability of two-dimensional time-harmonic electromagnetic problems involving non-conducting moving objects with stationary boundaries. *ESAIM Math. Model. Numer. Anal.* **2015**, *49*, 1157–1192. [CrossRef]
7. Ioannidis, A.D.; Kristensson, G.; Stratis, I.G. On the well-posedness of the Maxwell system for linear bianisotropic media. *SIAM J. Math. Anal.* **2012**, *44*, 2459–2473. [CrossRef]
8. Cocquet, P.; Mazet, P.; Mouysset, V. On the existence and uniqueness of a solution for some frequency-dependent partial differential equations coming from the modeling of metamaterials. *SIAM J. Math. Anal.* **2012**, *44*, 3806–3833. [CrossRef]
9. Kalarickel Ramakrishnan, P.; Raffetto, M. Well posedness and finite element approximability of three-dimensional time-harmonic electromagnetic problems involving rotating axisymmetric objects. *Symmetry* **2020**, *12*, 218. [CrossRef]
10. Wu, T.X.; Jaggard, D.L. A comprehensive study of discontinuities in chirowaveguides. *IEEE Trans. Microw. Theory Tech.* **2002**, *50*, 2320–2330. [CrossRef]
11. Alotto, P.; Codecasa, L. A fit formulation of bianisotropic materials over polyhedral grids. *IEEE Trans. Magn.* **2014**, *50*, 349–352. [CrossRef]
12. Monk, P. *Finite Element Methods for Maxwell's Equations*; Oxford Science Publications: Oxford, UK, 2003.
13. Fernandes, P.; Ottonello, M.; Raffetto, M. Regularity of time-harmonic electromagnetic fields in the interior of bianisotropic materials and metamaterials. *IMA J. Appl. Math.* **2014**, *79*, 54–93. [CrossRef]
14. Ciarlet, P.G.; Lions, J.L. *Handbook of Numerical Analysis*; Finite Element Methods, Part 1; North-Holland: Amsterdam, The Netherlands, 1991; Volume II.

Article

An Efficient Numerical Formulation for Wave Propagation in Magnetized Plasma Using PITD Method

Zhen Kang [1], Ming Huang [1], Weilin Li [1,*], Yufeng Wang [1] and Fang Yang [2]

[1] Department of Electrical Engineering, School of Automation, Northwestern Polytechnical University, Xi'an 710072, China; kangzhen@nwpu.edu.cn (Z.K.); minghuang@nwpu.edu.cn (M.H.); wyfnwpu@mail.nwpu.edu.cn (Y.W.)

[2] School of Electrical and Control Engineering, Xi'an University of Science and Technology, Xi'an 710054, China; yangf@xust.edu.cn

* Correspondence: li.weilin@hotmail.com; Tel.: +86-1357-2934-783

Received: 20 August 2020; Accepted: 24 September 2020; Published: 26 September 2020

Abstract: A modified precise-integration time-domain (PITD) formulation is presented to model the wave propagation in magnetized plasma based on the auxiliary differential equation (ADE). The most prominent advantage of this algorithm is using a time-step size which is larger than the maximum value of the Courant–Friedrich–Levy (CFL) condition to achieve the simulation with a satisfying accuracy. In this formulation, Maxwell's equations in magnetized plasma are obtained by using the auxiliary variables and equations. Then, the spatial derivative is approximated by the second-order finite-difference method only, and the precise integration (PI) scheme is used to solve the resulting ordinary differential equations (ODEs). The numerical stability and dispersion error of this modified method are discussed in detail in magnetized plasma. The stability analysis validates that the simulated time-step size of this method can be chosen much larger than that of the CFL condition in the finite-difference time-domain (FDTD) simulations. According to the numerical dispersion analysis, the range of the relative error in this method is 10^{-6} to 5×10^{-4} when the electromagnetic wave frequency is from 1 GHz to 100 GHz. More particularly, it should be emphasized that the numerical dispersion error is almost invariant under different time-step sizes which is similar to the conventional PITD method in the free space. This means that with the increase of the time-step size, the presented method still has a lower computational error in the simulations. Numerical experiments verify that the presented method is reliable and efficient for the magnetized plasma problems. Compared with the formulations based on the FDTD method, e.g., the ADE-FDTD method and the JE convolution FDTD (JEC-FDTD) method, the modified algorithm in this paper can employ a larger time step and has simpler iterative formulas so as to reduce the execution time. Moreover, it is found that the presented method is more accurate than the methods based on the FDTD scheme, especially in the high frequency range, according to the results of the magnetized plasma slab. In conclusion, the presented method is efficient and accurate for simulating the wave propagation in magnetized plasma.

Keywords: auxiliary differential equation (ADE); magnetized plasma; numerical simulation; PITD method; propagation

---

1. Introduction

The simulations of the electromagnetic (EM) wave propagation in the magnetized plasma are attractive and have a wide range of applications, e.g., high frequency components, PCB design, microstrip antenna, and so on [1–6]. Recently, the finite-difference time-domain (FDTD) formulation is the most popular numerical tool in the full wave analysis, and widely used in the simulation of the magnetized plasma and other dispersive materials. The typical algorithms based on the FDTD method for modeling the dispersive material include the recursive convolution (RC) FDTD method [7,8], the auxiliary differential equation (ADE) FDTD method [9–11], and the Z-transform (ZT) FDTD method [12,13]. Unfortunately, the FDTD method has an inherent drawback on which the above methods are based, i.e., Courant–Friedrich–Levy (CFL) stability criterion, and it has limited the further applications of the FDTD method in the dispersive materials as the problem expands. Assume that the spatial mesh is very fine to obtain the satisfying accuracy, the CFL stability criterion indicates that the time-step size should be a small enough value to ensure the numerical results are convergent. It is easily seen that such a small time-step size leads to an increase of the iteration step and a higher computational memory requirement so as to decrease the effectiveness of the FDTD calculation significantly and sometimes increase the accumulated error of the simulation.

Prompted by the aforementioned reasons, a number of time-domain algorithms, which have looser stability conditions, are presented to improve the efficiency of the FDTD simulations, e.g., the alternating-direction implicit FDTD (ADI-FDTD) method [14,15], the locally-one-dimensional (LOD) FDTD method [16–18], the precise-integration time-domain (PITD) method [19], and so on. Therein, the ADI-FDTD method is established by the alternating-direction implicit technique, and it is implicit and has complex iterative formulas. The LOD-FDTD method, which is presented by Shibayama, et al., is based on both the locally-one-dimensional scheme and the split time step technique. Compared with the ADI-FDTD method, the LOD-FDTD method is more efficient because it has fewer arithmetical operations. Here, we know both the ADI-FDTD and LOD-FDTD methods are unconditional stable to solve the EM wave problems. This means that the CFL condition has no restraint on the time step, and the efficiency of the simulation can be improved significantly by employing a larger time step. Meanwhile, the numerical dispersion errors of the ADI-FDTD method and the LOD-FDTD method are comparable and similar. It should be emphasized that the computational errors of these two methods increase rapidly which leads to the unanticipated results when the time step is increased as shown in the following description. In recent years, the ADI-FDTD method and the LOD-FDTD method were also generalized to the application of the dispersive materials [20,21].

The PITD method, which is presented by Ma, et al., has been used to solve Maxwell's equations in free space, lossy space, and unmagnetized plasma [19,22,23]. Although the PITD method is not unconditionally stable, it can still employ a time step which is much larger than the maximum value allowed by the CFL stability condition in the FDTD method [24]. Moreover, in contrast to the two unconditional stable methods mentioned above, the dispersion error of the PITD method is almost invariant when the time step is increased. This means that the PITD method can maintain a satisfactory computational accuracy when the time step is increased [25,26], and it makes the PITD method more suitable for the electrically large EM problems. Here, we take the ADI-FDTD method as the example. Figure 1 shows the numerical velocity of the FDTD, PITD, and ADI-FDTD methods with respect to the propagation angle when the Courant number is $S = 0.5$. It is found that the numerical dispersion error of the PITD method is larger than that of the FDTD method, but smaller than that of the ADI-FDTD method. Figure 2 shows the numerical velocity of the PITD method and the ADI-FDTD method versus the propagation angle under different Courant numbers. With the increase of the Courant number, the numerical dispersion error of the ADI-FDTD method is increased rapidly, however, the numerical dispersion error of the PITD method is still nearly invariant.

Due to the advantages mentioned above, the PITD method is a promising approach to model the EM wave propagation in magnetized plasma efficiently. The resulting Maxwell's equations of magnetized plasma are firstly obtained by employing the auxiliary variables and equations related to the current density. Then, the second-order accurate finite-difference formulation is used to approximate the spatial derivative in the presented method, and several ordinary differential equations (ODEs) with respect to the time derivative are obtained directly. Finally, we use the PI scheme to solve the ODEs. After establishing the modified PITD method in magnetized plasma, the stability condition and the dispersion error are analyzed numerically. The stability analysis verifies that the numerical stability criterion of the presented method in magnetized plasma is much looser than the CFL stability condition of the FDTD methods so as to increase the maximum allowable time step, and the numerical dispersion errors are almost invariant when the time-step size is increased. Thus, this method has the potential to balance both the efficiency and the accuracy. The magnetized plasma slab and the magnetized plasma filled cavity are simulated to validate that the modified PITD method in the paper is reliable and efficient. The analyses of the numerical results indicate that the presented method can provide an evident reduction of the execution time by using a larger simulated time step, meanwhile, the computational error of the presented algorithm is also lower than those of the formulations based on the FDTD scheme, especially in the high frequency range.

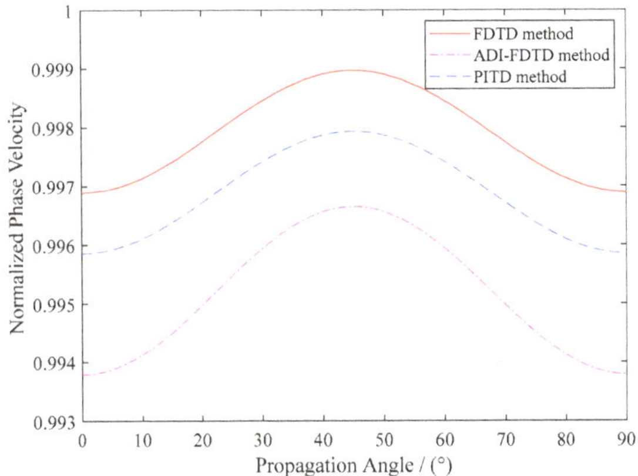

Figure 1. Numerical velocity of the finite-difference time-domain (FDTD), precise-integration time-domain (PITD) and alternating-direction implicit FDTD (ADI-FDTD) methods versus the propagation angle when the Courant number is $S = 0.5$.

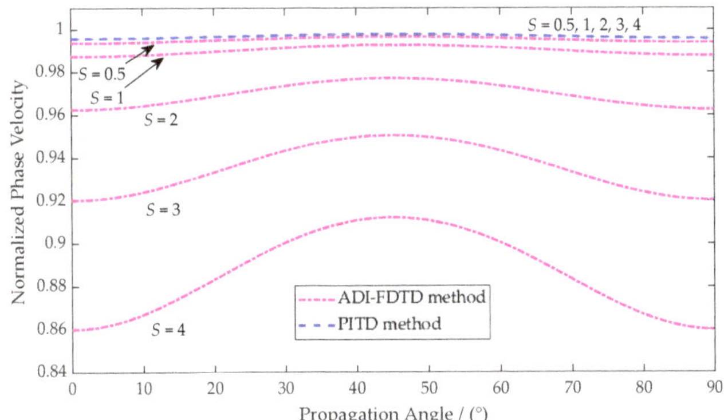

Figure 2. Numerical velocity of the PITD and ADI-FDTD methods versus the propagation angle under different Courant numbers.

The rest of this paper is organized as follows. The formulation of the presented method is introduced in Section 2. The stability analysis and the dispersion analysis are discussed numerically in Section 3. Numerical results are simulated to verify the efficiency and the accuracy of the presented algorithm in Section 4. Finally, conclusions are drawn in Section 5.

2. Formulations

2.1. Resulting Maxwell's Equations of Magnetized Plasma

The curl Maxwell's equations for describing the magnetized plasma problem is firstly established by employing the auxiliary variables and equations related to the current density $\mathbf{J}(t)$. The resulting matrix form is shown as follows:

$$\frac{\partial}{\partial t}\begin{bmatrix} \mathbf{E}(t) \\ \mathbf{H}(t) \\ \mathbf{J}(t) \end{bmatrix} = \begin{bmatrix} 0 & \frac{1}{\varepsilon_0}\nabla\times & -\frac{1}{\varepsilon_0} \\ -\frac{1}{\mu_0}\nabla\times & 0 & 0 \\ \varepsilon_0\omega_p^2 & 0 & -\gamma+\boldsymbol{\omega}_b\times \end{bmatrix}\begin{bmatrix} \mathbf{E}(t) \\ \mathbf{H}(t) \\ \mathbf{J}(t) \end{bmatrix}, \tag{1}$$

where ω_p and γ are the natural angular frequency and the collision frequency, respectively; $\boldsymbol{\omega}_b = e\mathbf{B}_0/m$ is the electron cyclotron angular frequency, wherein $\mathbf{B}_0$ is the applied magnetic field, e is the electric quantity, and m is the mass of the electron.

Assume that the applied magnetic field is z-direction in the following analysis, and Equation (1) can be expanded as follows:

$$\frac{\partial}{\partial t}\begin{bmatrix} E_x \\ E_y \\ E_z \\ H_x \\ H_y \\ H_z \\ J_x \\ J_y \\ J_z \end{bmatrix} = \begin{bmatrix} 0 & 0 & 0 & 0 & -\frac{1}{\varepsilon_0}\frac{\partial}{\partial z} & \frac{1}{\varepsilon_0}\frac{\partial}{\partial y} & -\frac{1}{\varepsilon_0} & 0 & 0 \\ 0 & 0 & 0 & \frac{1}{\varepsilon_0}\frac{\partial}{\partial z} & 0 & -\frac{1}{\varepsilon_0}\frac{\partial}{\partial x} & 0 & -\frac{1}{\varepsilon_0} & 0 \\ 0 & 0 & 0 & -\frac{1}{\varepsilon_0}\frac{\partial}{\partial y} & \frac{1}{\varepsilon_0}\frac{\partial}{\partial x} & 0 & 0 & 0 & -\frac{1}{\varepsilon_0} \\ 0 & \frac{1}{\mu_0}\frac{\partial}{\partial z} & -\frac{1}{\mu_0}\frac{\partial}{\partial y} & 0 & 0 & 0 & 0 & 0 & 0 \\ -\frac{1}{\mu_0}\frac{\partial}{\partial z} & 0 & \frac{1}{\mu_0}\frac{\partial}{\partial x} & 0 & 0 & 0 & 0 & 0 & 0 \\ \frac{1}{\mu_0}\frac{\partial}{\partial y} & -\frac{1}{\mu_0}\frac{\partial}{\partial x} & 0 & 0 & 0 & 0 & 0 & 0 & 0 \\ \varepsilon_0\omega_p^2 & 0 & 0 & 0 & 0 & 0 & -\gamma & -\omega_b & 0 \\ 0 & \varepsilon_0\omega_p^2 & 0 & 0 & 0 & 0 & \omega_b & -\gamma & 0 \\ 0 & 0 & \varepsilon_0\omega_p^2 & 0 & 0 & 0 & 0 & 0 & -\gamma \end{bmatrix}\begin{bmatrix} E_x \\ E_y \\ E_z \\ H_x \\ H_y \\ H_z \\ J_x \\ J_y \\ J_z \end{bmatrix}. \tag{2}$$

Here, we know that in the formulations based on the FDTD scheme, both the spatial and time derivatives are discretized by using the finite difference technique to obtain a set of algebraic equations from the Maxwell's equations. However, in the proposed PITD algorithm, only the spatial derivative is discretized, and several ODEs are obtained temporarily and written as matrix form:

$$\frac{d\mathbf{Y}}{dt} = \mathbf{H}\mathbf{Y} + \mathbf{g}(t), \tag{3}$$

where $\mathbf{Y} = \begin{bmatrix} E_x, E_y, E_z, H_x, H_y, H_z, J_x, J_y, J_z \end{bmatrix}^{\mathrm{T}}$, the coefficient matrix $\mathbf{H}$ is determined by both the EM parameters of the medium and the spatial step of the simulation, and $\mathbf{g}(t)$ is an inhomogeneous term related to the excitations.

The analytical and discrete solutions of Equation (3) are:

$$\mathbf{Y}(t) = e^{\mathbf{H}t}\mathbf{Y}(0) + \int_0^t e^{\mathbf{H}(t-\tau)}\mathbf{g}(\tau)d\tau, \tag{4}$$

and

$$\mathbf{Y}_{n+1} = \mathbf{T}\mathbf{Y}_n + \mathbf{T}^{n+1}\int_{t_n}^{t_{n+1}} e^{-\tau\mathbf{H}}\mathbf{g}(\tau)d\tau, \tag{5}$$

where $\mathbf{Y}_n = \mathbf{Y}(n\Delta t)$ is the discrete form of $\mathbf{Y}(t)$ and $\mathbf{T} = \exp(\mathbf{H}\Delta t)$ which can be calculated by the the PI technique.

2.2. PI Technique Review

The exponential matrix $\mathbf{T}$ is firstly reformulated as:

$$\mathbf{T} = [\exp(\mathbf{H}s)]^{l}, \tag{6}$$

where $s = \Delta t/l$, $l = 2^{N}$, and N is a preselected arbitrary integer. If N is large enough, the interval of s will be extremely small. Then, the Taylor series expansion is employed to approximate $\exp(\mathbf{H}s)$ with a high precision as shown in the following:

$$\exp(\mathbf{H}s) \approx \mathbf{I} + \mathbf{T}_{a}, \tag{7}$$

where:

$$\mathbf{T}_{a} = (\mathbf{H}s) + \frac{(\mathbf{H}s)^{2}}{2!} + \frac{(\mathbf{H}s)^{3}}{3!} + \frac{(\mathbf{H}s)^{4}}{4!}, \tag{8}$$

and evidently,

$$\mathbf{T} \approx (\mathbf{I} + \mathbf{T}_{a})^{l}. \tag{9}$$

It should be noted that if $\mathbf{T}_{a}$ is added to the identity matrix $\mathbf{I}$ directly, $\mathbf{T}_{a}$ will be neglected because $\mathbf{T}_{a}$ is extremely small, which leads to a precision reduction of the exponential matrix. Therefore, it is evident that $\mathbf{T}_{a}$ should be operated in the process.

The matrix $\mathbf{T}$ is computed as follows:

$$\mathbf{T} \approx (\mathbf{I} + \mathbf{T}_{a})^{l} = (\mathbf{I} + \mathbf{T}_{a})^{2^{N}} = (\mathbf{I} + \mathbf{T}_{a})^{2^{N-1}} \times (\mathbf{I} + \mathbf{T}_{a})^{2^{N-1}} = \cdots, \tag{10}$$

and

$$(\mathbf{I} + \mathbf{T}_{a})^{2} = \mathbf{I} + 2\mathbf{T}_{a} + \mathbf{T}_{a} \times \mathbf{T}_{a}. \tag{11}$$

Start with Equation (8) to compute $\mathbf{T}_{a}$ and then run the following instruction, the exponential matrix $\mathbf{T}$ can be calculated:

do. $i = 1, N$

$$\mathbf{T}_{a} = 2\mathbf{T}_{a} + \mathbf{T}_{a} \times \mathbf{T}_{a}. \tag{12}$$

end do

$$\mathbf{T} = \mathbf{I} + \mathbf{T}_{a}. \tag{13}$$

Equations (8), (12), and (13) constitute the whole process of the PI technique to calculate the exponential matrix. Relying on the previous work, the simulated time step of the PITD method is much larger than the maximum allowable value of the CFL stability condition of the FDTD formulation in the lossless or lossy problems, which is very significant in the full wave analysis. For the magnetized plasma material, we believe the application of the PI technique can achieve the same effect and the following stability analysis and numerical experimentations will prove this point.

Furthermore, for the integration on the right-hand side of Equation (5), a linear variation of the term $\mathbf{g}(t)$ is assumed within the interval (t_{n}, t_{n+1}), expressed as:

$$\mathbf{g}(t) = \mathbf{r}_{0} + \mathbf{r}_{1}(t - t_{n}), \tag{14}$$

Substitute the above expression in the integration, we have the following recursive form solution:

$$\mathbf{Y}_{n+1} = \mathbf{T}\left[\mathbf{Y}_{n} + \mathbf{M}^{-1}(\mathbf{r}_{0} + \mathbf{M}^{-1}\mathbf{r}_{1})\right] - \mathbf{M}^{-1}\left[\mathbf{r}_{0} + \mathbf{M}^{-1}\mathbf{r}_{1} + \mathbf{r}_{1}\Delta t\right]. \tag{15}$$

In most cases, Equation (15) is unavailable directly since the matrix $\mathbf{M}$ is noninvertible. To mitigate this problem, the three-points Gauss integral formula is adopted to calculate the integration in Equation (5), and the recurrence formula is obtained as follows:

$$\mathbf{Y}_{n+1} = \mathbf{T}\mathbf{Y}_n + \tfrac{5\Delta t}{18}\exp\!\left[\left(\tfrac{\Delta t}{2} + \tfrac{\Delta t}{2}\sqrt{\tfrac{3}{5}}\right)\mathbf{H}\right]\mathbf{g}\!\left(t_n + \tfrac{\Delta t}{2} - \tfrac{\Delta t}{2}\sqrt{\tfrac{3}{5}}\right)$$

$$+\tfrac{8\Delta t}{18}\exp\!\left[\tfrac{\Delta t}{2}\mathbf{H}\right]\mathbf{g}\!\left(t_n + \tfrac{\Delta t}{2}\right) \tag{16}$$

$$+\tfrac{5\Delta t}{18}\exp\!\left[\left(\tfrac{\Delta t}{2} - \tfrac{\Delta t}{2}\sqrt{\tfrac{3}{5}}\right)\mathbf{H}\right]\mathbf{g}\!\left(t_n + \tfrac{\Delta t}{2} + \tfrac{\Delta t}{2}\sqrt{\tfrac{3}{5}}\right)$$

3. Stability Analysis and Numerical Dispersion Analysis

3.1. Stability Analysis

In this subsection, the amplitude of the eigenvalues of the exponential matrix $\mathbf{T}$ is used to discuss the numerical stability of the presented PITD algorithm in the magnetized plasma. Von Neumann stability criterion indicates that if the amplitudes of all the eigenvalues of the exponential matrix $\mathbf{T}$ are no larger than unity, the update equations of the presented PITD algorithm will be stable.

We consider the 2-D case in the following analysis. The preselected integer N is selected as 20, and $l = 2^{20}$ in the proposed method. The natural angular frequency of the magnetized plasma discussed is $2\pi \times 28.7 \times 10^9$ rad/s, the collision frequency is 20 GHz, and the electron cyclotron frequency is set as 1.0×10^{11} rad/s. Figure 3 graphs the comparison of the unit circle and the eigenvalues of the exponential matrix $\mathbf{T}$ in the complex plane when the time-step size is $10^6 \Delta t_{\text{CFL}}$. Here, Δt_{CFL} is the upper limit time-step size of the conventional FDTD method. It is clear seen that all the eigenvalues are within or on the unit circle, which means that the presented PITD method of the magnetized plasma is stable under such a large time-step size. Therefore, the proposed formulation can use a time-step size much larger than the maximum value of the CFL stability condition to achieve the simulation of the magnetized plasma problems.

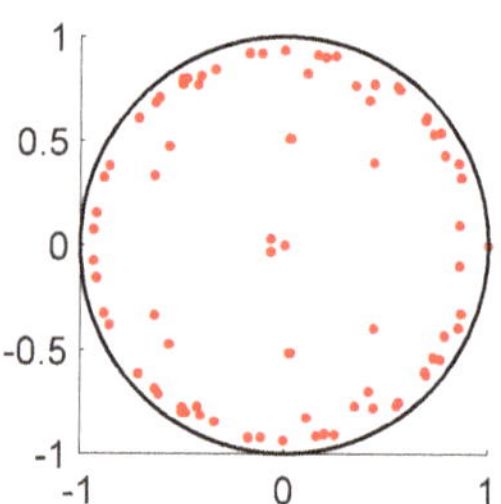

Figure 3. Comparison of the unit circle and the eigenvalues of the presented PITD algorithm in the complex plane.

3.2. Numerical Dispersion Analysis

In this subsection, the dispersion performance of the presented PITD method in magnetized plasma is discussed numerically by adopting the Fourier approach. The dispersion performance of the presented formulation is described by the differences between the numerical wave number k_{num} and the analytical wave number k_{anal}. Suppose c is the velocity of light in the vacuum, the analytic wave number of the left-hand circularly polarized (LCP) EM wave is:

$$k_L = \frac{\omega}{c}\sqrt{1 - \frac{\omega_p^2}{\omega^2\left[\left(1 - j\frac{\gamma}{\omega}\right) + \frac{\omega_b}{\omega}\right]}}, \tag{17}$$

and the analytic wave number of the right-hand circularly polarized (RCP) EM waves is:

$$k_R = \frac{\omega}{c}\sqrt{1 - \frac{\omega_p^2}{\omega^2\left[\left(1 - j\frac{\gamma}{\omega}\right) - \frac{\omega_b}{\omega}\right]}}. \tag{18}$$

Assuming that the monochromatic plane wave propagates in the magnetized plasma, the field components are expressed as:

$$\mathbf{X} = \mathbf{X}_0 e^{-j(kz - \omega t)},$$ (19)

where k is the wave number, $\mathbf{X}_0$ and ω are the amplitude and the angular frequency of the EM wave, respectively. The discrete form of Equation (19) is obtained as follows:

$$\mathbf{X}_k = \mathbf{X}_0 e^{-j(km\Delta z - \omega n \Delta t)},$$ (20)

where m and n are the space index and the time index, respectively.

Here, we consider the 1-D case, and the vector $\mathbf{X}$ includes the field components E_x, E_y, H_x, H_y, J_x, and J_y. Substituting the discrete form of the field components into Equation (2) for the 1-D case, a homogenous system of ODEs can be obtained and written in matrix form as:

$$\frac{d\mathbf{X}}{dt} = \mathbf{H}_1 \mathbf{X}.$$ (21)

Here, the coefficient matrix $\mathbf{H}_1$ is:

$$\mathbf{H}_1 = \begin{bmatrix} 0 & 0 & 0 & \frac{jW}{\varepsilon_0} & -\frac{1}{\varepsilon_0} & 0 \\ 0 & 0 & -\frac{jW}{\varepsilon_0} & 0 & 0 & -\frac{1}{\varepsilon_0} \\ 0 & -\frac{jW}{\mu_0} & 0 & 0 & 0 & 0 \\ \frac{jW}{\mu_0} & 0 & 0 & 0 & 0 & 0 \\ \varepsilon_0 \omega_p^2 & 0 & 0 & 0 & -\gamma & -\omega_b \\ 0 & \varepsilon_0 \omega_p^2 & 0 & 0 & \omega_b & -\gamma \end{bmatrix},$$ (22)

where:

$$W = \frac{2 \sin\left(\frac{1}{2} k \Delta z\right)}{\Delta z}.$$ (23)

The following discrete iterative formula is used to solve the ODEs Equation (21):

$$\mathbf{X}_{k+1} = \mathbf{T}_1 \mathbf{X}_k,$$ (24)

where the exponential matrix $\mathbf{T}_1$ is:

$$\mathbf{T}_1 = \left[\mathbf{I} + (\mathbf{H}_1 s) + \frac{(\mathbf{H}_1 s)^2}{2!} + \frac{(\mathbf{H}_2 s)^3}{3!} + \frac{(\mathbf{H}_3 s)^4}{4!} \right]^l.$$ (25)

Then, we have:

$$\left(e^{j\omega t} \mathbf{I} - \mathbf{T}_1 \right) \mathbf{X}_0 = 0.$$ (26)

Since it is true for any $\mathbf{X}_0$ that is nonzero, the determinant of the coefficient matrix $\left(e^{j\omega t} \mathbf{I} - \mathbf{T} \right)$ should be zero:

$$\left(e^{j\omega t} \mathbf{I} - \mathbf{T}_1 \right) = 0.$$ (27)

In the following analysis, the numerical dispersion error and the numerical dissipation error are defined to describe the precision of the presented PITD method in the magnetized plasma. The definition of the relative dispersion error is $(\mathrm{Re}(k_{num}) - \mathrm{Re}(k_{anal}))/\mathrm{Re}(k_{anal})$, and it is related to the phase error. The definition of the relative dissipation error is $(\mathrm{Im}(k_{num}) - \mathrm{Im}(k_{anal}))/\mathrm{Im}(k_{anal})$, and it is related to the amplitude error [27].

It is assumed that $\Delta z = 75$ μm, $\Delta t = 0.125$ ps, and $\omega_b = 3.0 \times 10^{11}$ rad/s. The preselected integer N is chosen as 20, and $l = 2^{20}$ in the presented PITD method. The solutions of numerical wave number are computed by Equation (27).

3.2.1. Effect of Wave Frequency on Numerical Error

The natural angular frequency of the magnetized plasma is $\omega_p = 2\pi \times 50 \times 10^9$ rad/s. The collision frequency is 20 GHz. Figure 4a,b graphs the relative dispersion and relative dissipation errors with respect to the wave frequency ω for both the LCP and RCP waves, respectively. We can see that both the dispersion and dissipation errors increase monotonically with the wave frequency. Furthermore, the relative dispersion error is higher than the relative dissipation error in the LCP wave, and is lower than the relative dissipation error in the RCP wave. In addition, the relative error range of the proposed PITD method is 10^{-6} to 5×10^{-4} when the EM wave frequency is from 1 GHz to 100 GHz.

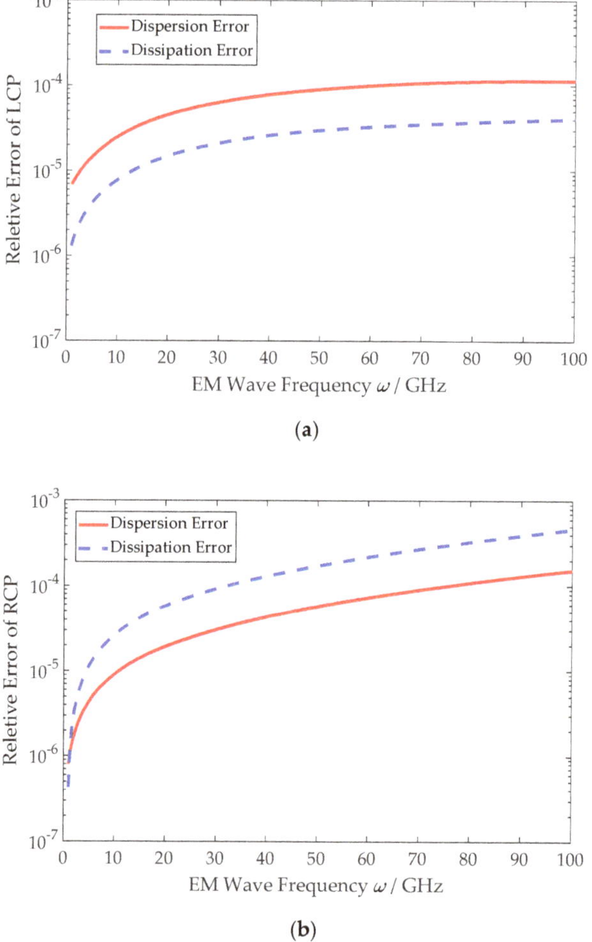

Figure 4. Relative dispersion and relative dissipation errors with respect to the electromagnetic (EM) wave frequency: (**a**) The left-hand circularly polarized (LCP) wave. (**b**) The right-hand circularly polarized (RCP) wave.

3.2.2. Effect of the Natural Angular Frequency on Numerical Error

The EM wave frequency is set as 50 GHz. The collision frequency of the magnetized plasma is 20 GHz. Figure 5a,b graphs the relative dispersion and relative dissipation errors with respect to the natural angular frequency ω_p for both the LCP and RCP waves, respectively. For the LCP wave, the relative dispersion error curve has lower-peak error when $\omega_p/2\pi$ is 18 GHz, and the relative dissipation error curve has lower-peak errors when $\omega_p/2\pi$ are 4 GHz and 21 GHz. For the RCP wave, the relative dispersion error curve has no lower-error peak, and the relative dissipation error curve has lower-peak error when $\omega_p/2\pi$ is 6 GHz. Moreover, both the relative dispersion and relative dissipation errors increase monotonically with the natural frequency when $\omega_p/2\pi$ is larger than the frequency of the lower-peak error.

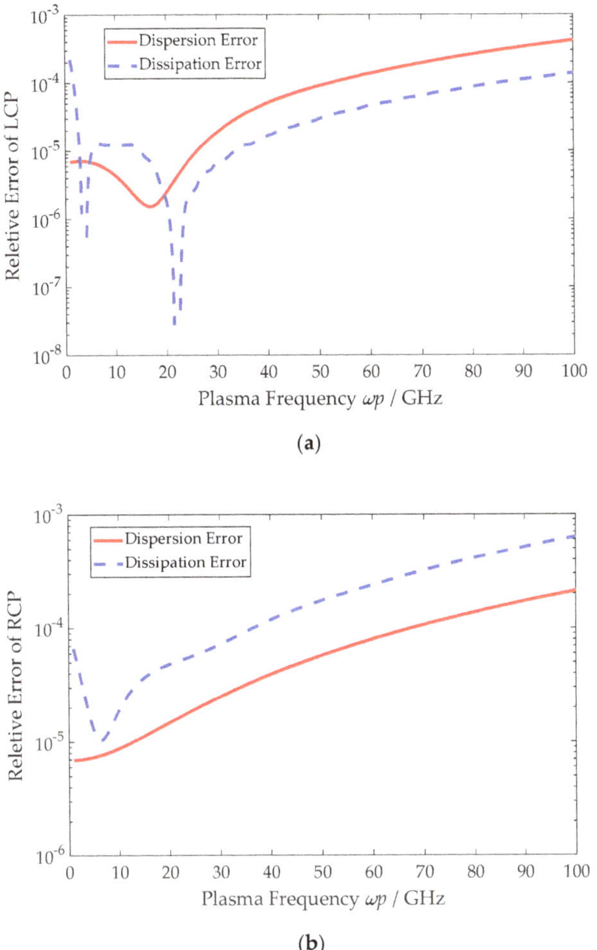

(a)

(b)

Figure 5. Relative dispersion and relative dissipation errors with respect to the natural frequency: (**a**) The LCP wave. (**b**) The RCP wave.

3.2.3. Effect of the Plasma Collision Frequency on Numerical Error

The EM wave frequency is set as 50 GHz, and the natural angular frequency is $\omega_p = 2\pi \times 50 \times 10^9$ rad/s. Figure 6a,b graphs the relative dispersion and relative dissipation errors with respect to the collision frequency γ for both the LCP and RCP waves, respectively. It is found that the relative dispersion and relative dissipation errors are slightly decreased when the collision frequency is increased.

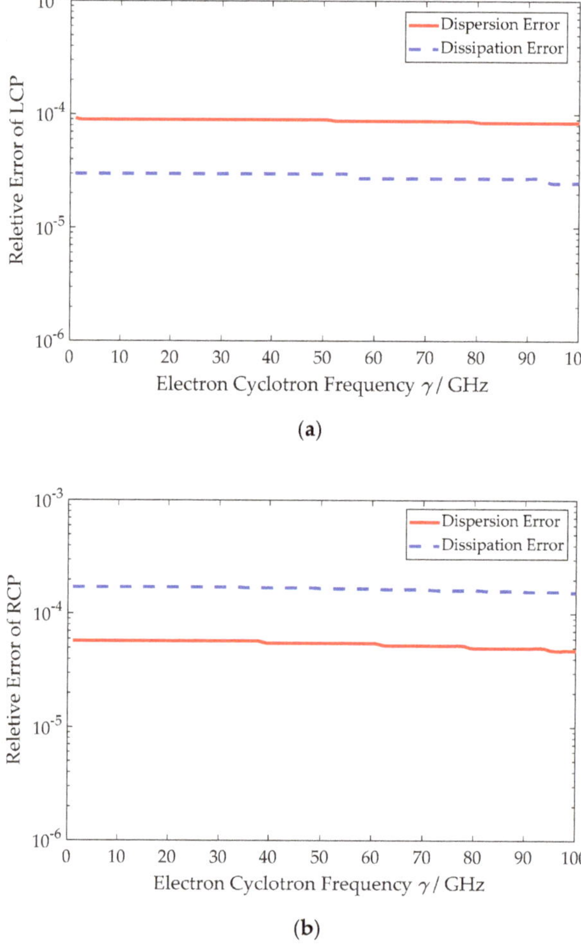

(a)

(b)

Figure 6. Relative dispersion and relative dissipation errors with respect to the collision frequency: (a) The LCP wave. (b) The RCP wave.

3.2.4. Effect of the Time-Step Size on Numerical Error

It is assumed that the natural angular frequency is $2\pi \times 50 \times 10^9$ rad/s, and the collision frequency is 20 GHz. Figure 7 graphs the relative dispersion and relative dissipation errors with respect to the wave frequency ω for both the LCP and RCP waves under different Courant number S, respectively. It is clear that all the curves are in agreement. Figure 8 shows the relative dispersion and relative dissipation errors with respect to the Courant number when the EM wave frequency is 50 GHz. Figure 8 indicates that the relative dispersion and relative dissipation errors are almost invariant when the Courant numbers is increased. These mean that the proposed method can maintain a lower numerical dispersion error in the simulations when the time step is increased.

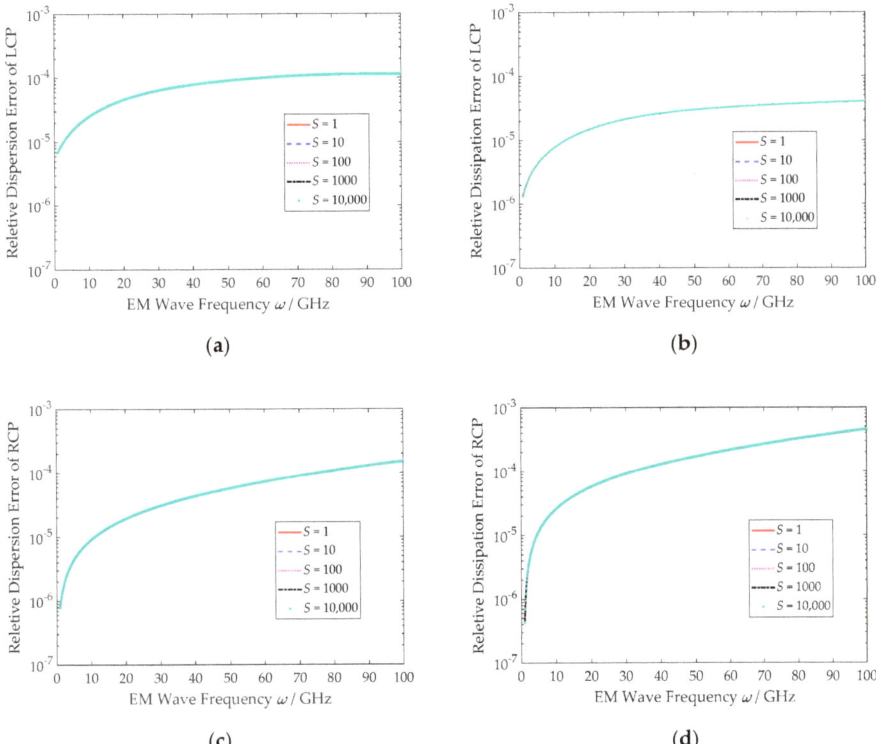

Figure 7. Relative dispersion and relative dissipation errors with respect to the wave frequency under different Courant numbers: (**a**) Relative dispersion error of the LCP wave. (**b**) Relative dissipation error of the LCP wave. (**c**) Relative dispersion error of the RCP wave. (**d**) Relative dissipation error of the RCP wave.

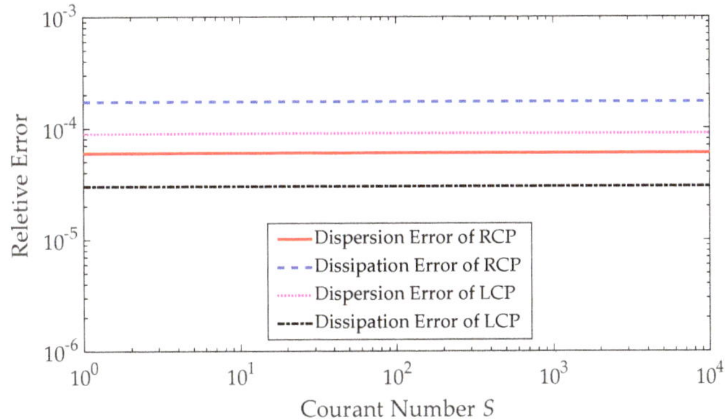

Figure 8. Relative dispersion and relative dissipation errors with respect to the Courant numbers for the LCP wave and the RCP wave.

4. Numerical Experiments

The performance of the presented PITD method are verified by two typical magnetized plasma examples which are also solved by the analytical formulas and the formulations based on the FDTD method, respectively, for comparison.

4.1. Magnetized Plasma Slab

As the first example, a magnetized plasma slab is simulated to validate the efficiency and the accuracy of the modified PITD algorithm in this paper. The diagram of the infinite magnetized plasma slab in infinite free space is shown in Figure 9. The numerical reflection coefficient and transmission coefficient are computed by the presented method, the JEC-FDTD method and the ADE-FDTD method, respectively. The results are also compared with the analytical solution.

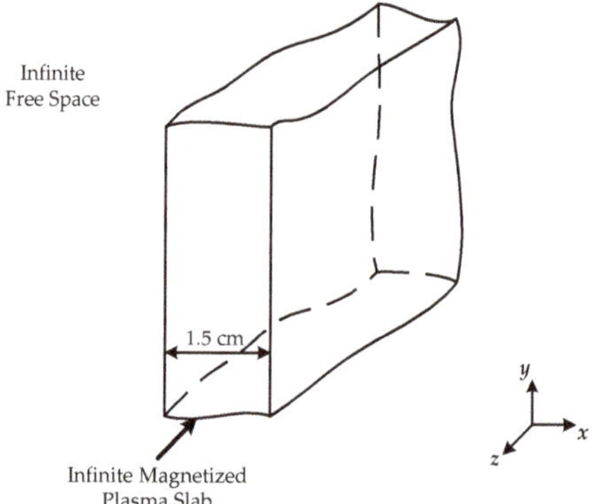

Figure 9. The diagram of the magnetized plasma slab in free space.

The magnetized plasma slab is 1.5 cm thick, and divided into 200 cells, i.e., the space step is 75 μm. The main computing region is composed of 600 free space cells (the space indexes are from 1 to 300, and 501 to 800) and 200 magnetized plasma cells (the space indexes are from 301 to 500). Perfectly matched layer (PML) is employed as the absorption boundary condition to eliminate the reflection error. The time-step size of the methods based on the FDTD formulation is set to $\Delta t_{\text{FDTD}} = 0.125$ ps, and the time-step size of the proposed PITD method is 5 times of Δt_{FDTD} (i.e., $\Delta t_{\text{ProposedPITD}} = 0.625$ ps). The parameters of the magnetized plasma are shown as follows:

$$\omega_p = 2\pi \times 28.7 \times 10^9 \text{rad/s}, \tag{28}$$

$$\gamma = 20.0 \text{ GHz}, \tag{29}$$

$$\omega_b = 1.0 \times 10^{11} \text{ rad/s}. \tag{30}$$

Figures 10–13 show the magnitude and the phase of the complex reflection coefficient and transmission coefficient of the magnetized plasma slab calculated by the JEC-FDTD method, the ADE-FDTD method, the proposed PITD method, and the analytical solution, respectively. We can clearly see that the computational results of the proposed PITD method are coincident with the analytical solutions on both the magnitude and the phase.

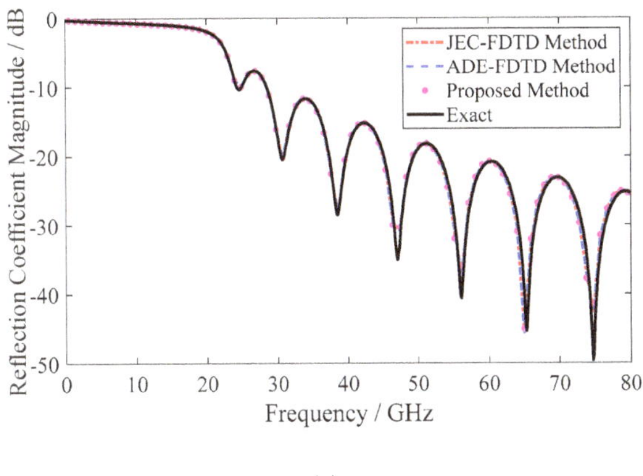

(**a**)

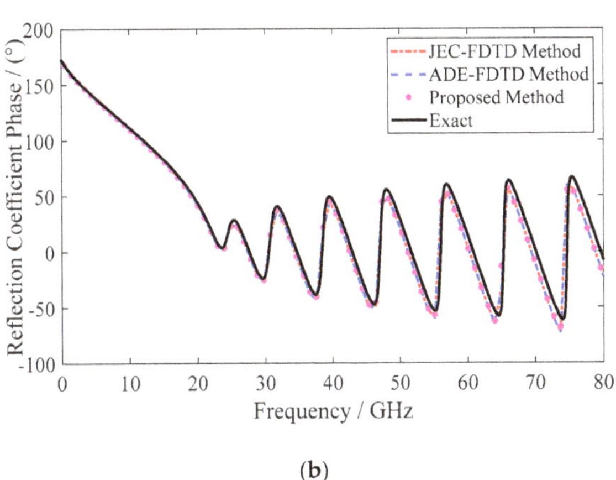

(**b**)

Figure 10. Calculated complex reflection coefficient of the LCP wave: (**a**) Magnitude. (**b**) Phase.

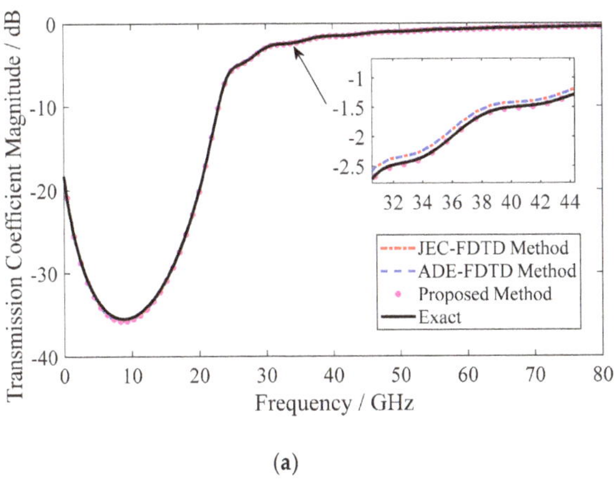

(**a**)

Figure 11. *Cont.*

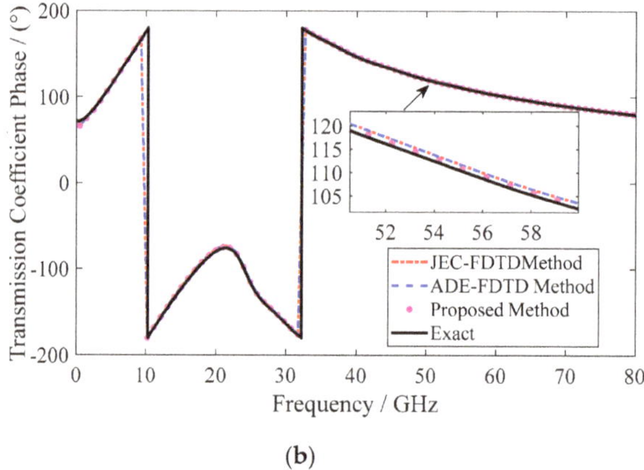

(**b**)

Figure 11. Calculated complex transmission coefficient of the LCP wave: (**a**) Magnitude. (**b**) Phase.

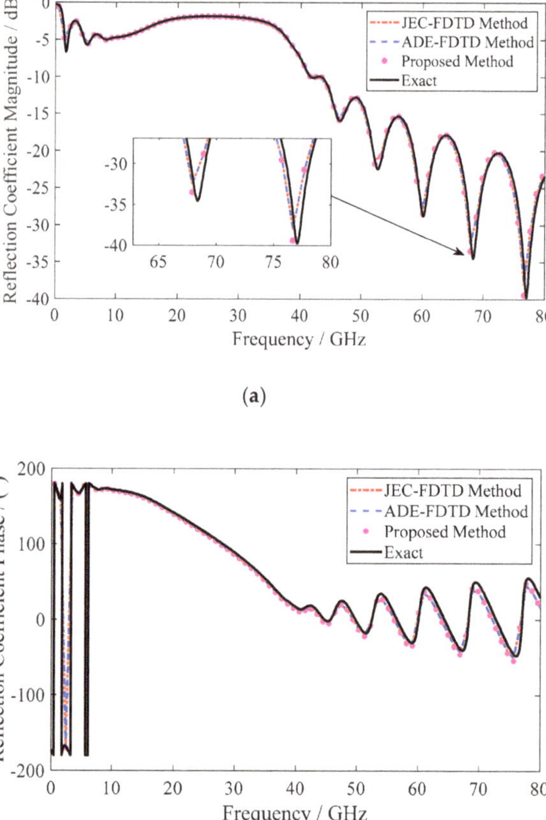

(**a**)

(**b**)

Figure 12. Calculated complex reflection coefficient of the RCP wave: (**a**) Magnitude. (**b**) Phase.

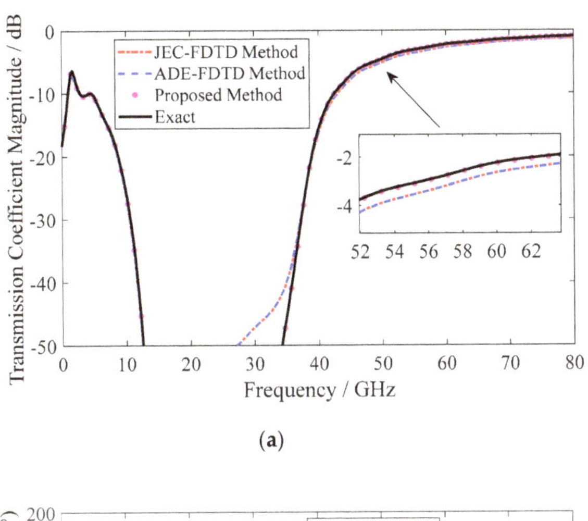

(a)

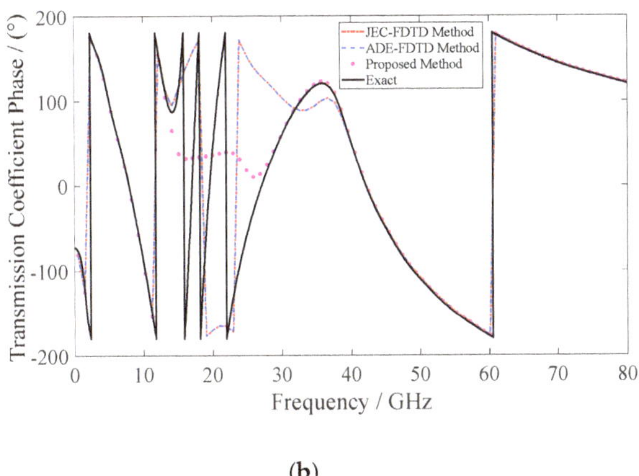

(b)

Figure 13. Calculated complex transmission coefficient of the RCP wave: (**a**) Magnitude. (**b**) Phase.

According to Figures 11, 12a and 13, it should be noted that the solutions of the proposed PITD method is more accurate than those of the JEC-FDTD method and the ADE-FDTD method, especially in the higher frequency range. Meanwhile, it is also found that larger errors occur in the stopband of the transmission coefficient for the RCP wave. Figure 13 shows that larger errors occur from 13 GHz to 38 GHz for the FDTD methods, and from 13 GHz to 27 GHz for the proposed method. This means that the low computational error range of the presented PITD method is larger than the FDTD methods.

The CPU time of the three methods are also recorded. The execution time of the JEC-FDTD method, the ADE-FDTD method, and the presented method are 8.50 s, 8.09 s, and 4.52 s, respectively. It can be concluded that a larger simulated time step leads to an appreciably reduction of the CPU time.

4.2. 2-D Magnetized Plasma Filled Cavity

The second example is a 2-D cavity filled with the magnetized plasma as shown in Figure 14. The main computing region is divided into 20×20 cells with a space step 75 μm. The time-step size of the ADE-FDTD method is $\Delta t = 0.1$ ps. For the proposed PITD method, the time-step size is $6\Delta t$ in this example. Therefore, the simulations are executed by 3000 iterative steps in the ADE-FDTD method and 500 iterative steps in the presented PITD method. The parameters of the magnetized plasma filled in the cavity are $\omega_p = 2\pi \times 28.7 \times 10^9$ rad/s, $\gamma = 10.0$ GHz, and $\omega_b = 3.0 \times 10^{11}$ rad/s.

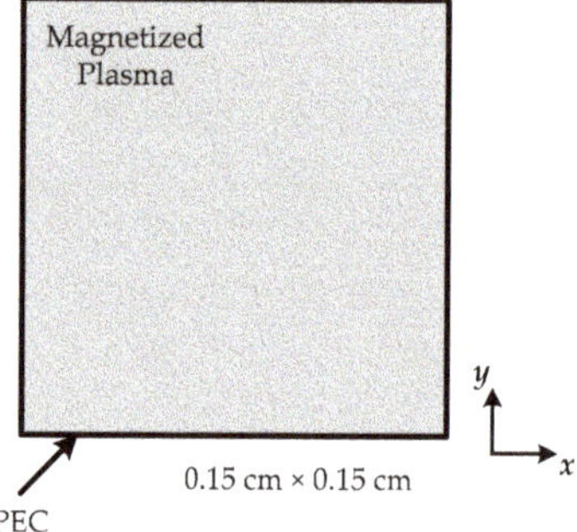

Figure 14. The diagram of the 2-D cavity filled with the magnetized plasma.

Figure 15 graphs the time-domain waveforms of the electric field E_x simulated by the presented PITD method and the ADE-FDTD method, respectively. It is shown that good agreements are achieved between the two methods on the simulated waveform. Table 1 lists the numerical resonant frequencies and the execution time of the presented PITD method and the ADE-FDTD method, respectively. It can be found that the calculated resonant frequencies are also coincident, moreover, the CPU time of the presented method is at least 1/3 less than that of the ADE-FDTD method. The simulations of both the FDTD methods and the PITD method in above analysis are achieved by MATLAB and performed on Intel(R) Core(TM) i3 CPU M370 2.40 GHz PC (Intel, Santa Clara, CA, USA).

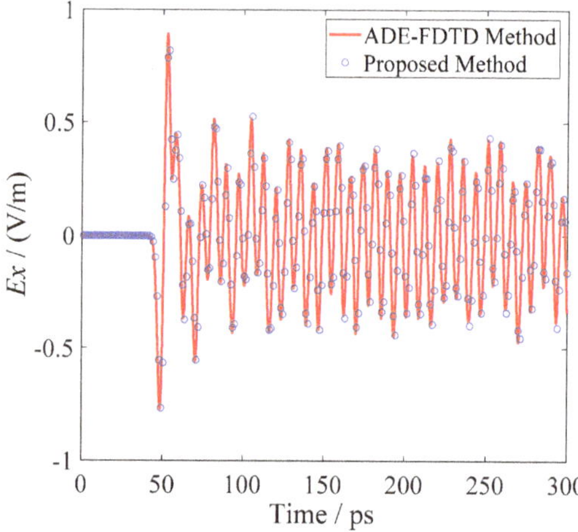

Figure 15. The time-domain waveforms of the electric field E_x simulated by the proposed PITD method and the auxiliary differential equation FDTD (ADE-FDTD) method.

Table 1. A comparison of calculated frequencies and execution time between the auxiliary differential equation finite-difference time-domain (ADE-FDTD) method and the proposed precise-integration time-domain (PITD) method.

Method	Resonant Frequency/GHz			Execution Time/s
ADE-FDTD	30.76	103.5	201.7	12.85
Proposed PITD	31.25	103.5	201.2	8.07

In conclusion, according to the numerical experiments above, the efficiency of the modified PITD method in this paper is higher than the algorithms based on the FDTD scheme for modeling the magnetized plasma. Meanwhile, the solutions of the magnetized plasma slab validate that the presented method is more accurate than the JEC-FDTD method and the ADE-FDTD method, especially in the high frequency range.

5. Conclusions

Based on both the auxiliary differential equation and the PI technique, a modified PITD method has been proposed for modeling the EM wave propagation through magnetized plasma in this paper. The analyses of the numerical stability and dispersion are discussed respectively, and the superior performance of the proposed method has been confirmed numerically. It is found that the numerical stability criterion of the proposed method is much looser than the CFL stability condition of the FDTD methods so as to increase the allowable simulated time step, and the numerical dispersion error and the dissipation error are almost invariant when the time step is increased. The numerical results validate the efficiency and accuracy of the presented algorithm. In the numerical experiments above, the proposed method can use a time-step size much larger than the value allowed by the CFL limit of the FDTD method which leads to a reduction of the CPU time in the simulation. Meanwhile, the accuracy performance of the presented PITD method is better than the JEC-FDTD method and the ADE-FDTD method, especially in higher frequency range. In conclusion, the proposed algorithm is a strong numerical tool to solve the EM wave problems in the magnetized plasma.

Author Contributions: Conceptualization, Z.K.; methodology, Z.K.; software, Z.K. and Y.W.; validation, M.H. and F.Y.; formal analysis, M.H. and Y.W.; investigation, F.Y.; resources, W.L.; data curation, M.H. and F.Y.; writing—original draft preparation, Z.K.; writing—review and editing, M.H., W.L., Y.W. and F.Y.; visualization, M.H. and F.Y.; supervision, W.L.; project administration, Z.K. and W.L.; funding acquisition, Z.K., W.L. and F.Y. All authors have read and agreed to the published version of the manuscript.

Funding: This research was supported by the China Postdoctoral Science Foundation under Grant number 2019M653737, and in part by the National Natural Science Foundation of Shaanxi Province under Grant number 2019JQ-226 and 2019JQ-792.

Conflicts of Interest: The authors declare no conflict of interest.

References

1. Letsholathebe, D.; Mphale, K. Microwave phase perturbation and ionisation measurement in vegetation fire plasma. *IET Microw. Antennas Propag.* **2013**, *7*, 741–745. [CrossRef]
2. Nadal, C.; Pigache, F.; Lefevre, Y. First approach for the modelling of the electric field surrounding a piezoelectric transformer in view of plasma generation. *IEEE Trans. Magn.* **2012**, *48*, 423–426. [CrossRef]
3. Li, X.S.; Hu, B.J. FDTD Analysis of a magneto-plasma antenna with uniform or nonuniform distribution. *IEEE Antennas Wirel. Propag. Lett.* **2010**, *9*, 175–178. [CrossRef]
4. Chai, S.W.; Lim, S.M.; Hong, S.C. THz detector with an antenna coupled stacked CMOS plasma-wave FET. *IEEE Microw. Wirel. Compon. Lett.* **2014**, *24*, 869–871. [CrossRef]
5. Vyas, H.; Chaudhury, B. Computational investigation of power efficient plasma-based reconfigurable microstrip antenna. *IET Microw. Antennas Propag.* **2018**, *12*, 1587–1593. [CrossRef]
6. Krishna, T.N.V.; Sathishkumar, P.; Himasree, P.; Punnoose, D.; Raghavendra, K.V.G.; Naresh, B.; Rana, R.A.; Kim, H.J. 4T analog MOS control-high voltage high frequency (HVHF) plasma switching power supply for water purification in industrial applications. *Electronics* **2018**, *7*, 245. [CrossRef]
7. Liu, J.F.; Lv, H.; Zhao, Y.C.; Fang, Y.; Xi, X.L. Stochastic PLRC-FDTD method for modeling wave propagation in unmagnetized plasma. *IEEE Antennas Wirel. Propag. Lett.* **2018**, *17*, 1024–1028. [CrossRef]
8. Shibayama, J.; Takahashi, R.; Yamauchi, J.; Nakano, H. Frequency-dependent LOD-FDTD implementations for dispersive media. *Electron. Lett.* **2006**, *42*, 1084–1086. [CrossRef]
9. Ramadan, O. Stability-improved ADE-FDTD implementation of Drude dispersive models. *IEEE Antennas Wirel. Propag. Lett.* **2018**, *17*, 877–880. [CrossRef]

10. Ramadan, O. Efficient ADE-WE-PML formulations for scalar dispersive FDTD applications. *Electron. Lett.* **2013**, *49*, 157–158. [CrossRef]

11. Li, J.X.; Wu, P.Y.; Jiang, H.L. Implementation of higher order CNAD CFS-PML for truncating unmagnetised plasma. *IET Microw. Antennas Propag.* **2019**, *13*, 756–760. [CrossRef]

12. Li, J.; Dai, J. Z-Transform implementations of the CFS-PML. *IEEE Antennas Wirel. Propag. Lett.* **2006**, *5*, 410–413. [CrossRef]

13. Taflove, A.; Hagness, S.C. *Computational Electrodynamics: The Finite-Difference Time-Domain Method*; Artech House: Boston, MA, USA, 2005.

14. Namiki, T. A new FDTD algorithm based on alternating-direction implicit method. *IEEE Trans. Microw. Theory Tech.* **1999**, *47*, 2003–2007. [CrossRef]

15. Namiki, T. 3-D ADI-FDTD method-unconditionally stable time-domain algorithm for solving full vector Maxwell's equations. *IEEE Trans. Microw. Theory Tech.* **2000**, *48*, 1743–1748. [CrossRef]

16. Shibayama, J.; Muraki, M.; Yamauchi, J.; Nakano, H. Efficient implicit FDTD algorithm based on locally one-dimensional scheme. *Electron. Lett.* **2005**, *41*, 1046–1047. [CrossRef]

17. Ahmed, I.; Chua, E.K.; Li, E.P.; Chen, Z. Development of the three-dimensional unconditionally stable LOD-FDTD method. *IEEE Trans. Antennas Propag.* **2008**, *56*, 3596–3600. [CrossRef]

18. Liang, T.L.; Shao, W.; Shi, S.B.; Ou, H. Analysis of extraordinary optical transmission with periodic metallic gratings using ADE-LOD-FDTD method. *IEEE Photonics J.* **2016**, *8*, 7804710. [CrossRef]

19. Ma, X.K.; Zhao, X.T.; Zhao, Y.Z. A 3-D precise integration time-domain method without the restrains of the Courant-Friedrich-Levy stability condition for the numerical solution of Maxwell's equations. *IEEE Trans. Microw. Theory Tech.* **2006**, *54*, 3026–3037.

20. Song, W.J.; Zhang, H. RKADE-ADI FDTD applied to analysis of terahertz wave characteristics of plasma. *IET Microw. Antennas Propag.* **2017**, *11*, 1071–1077.

21. Liang, T.L.; Shao, W.; Shi, S.B. Complex-envelope ADE-LOD-FDTD for band gap analysis of plasma photonic crystals. *Appl. Comput. Electromagn. Soc. J.* **2018**, *33*, 443–449.

22. Sun, G.; Ma, X.K.; Bai, Z.M. Numerical stability and dispersion analysis of the precise-integration time-domain method in lossy media. *IEEE Trans. Microw. Theory Tech.* **2012**, *60*, 2723–2729. [CrossRef]

23. Kang, Z.; Li, W.L.; Wang, Y.F.; Huang, M.; Yang, F. A precise-integration time-domain formulation based on auxiliary differential equation for transient propagation in plasma. *IEEE Access* **2020**, *8*, 59741–59749. [CrossRef]

24. Jiang, L.L.; Chen, Z.Z.; Mao, J.F. On the numerical stability of the precise integration time-domain (PITD) method. *IEEE Microw. Wirel. Compon. Lett.* **2007**, *17*, 471–473. [CrossRef]

25. Kang, Z.; Ma, X.K.; Zhuansun, X. An efficient 2-D compact precise-integration time-domain method for longitudinally invariant waveguiding structures. *IEEE Trans. Microw. Theory Tech.* **2013**, *61*, 2535–2544. [CrossRef]

26. Bai, Z.M.; Ma, X.K.; Sun, G. A low-dispersion realization of precise integration time-domain method using a fourth-order accurate finite difference scheme. *IEEE Trans. Antennas Propag.* **2011**, *59*, 1311–1320. [CrossRef]

27. Zhong, S.Y.; Lai, Z.Q.; Liu, S.; Liu, S.B. The numerical dispersion relation and stability analysis of PLCDRC-FDTD method for anisotropic magnetized plasma. In Proceedings of the International Conference on Microwave and Millimeter Wave Technology, Nanjing, China, 21–24 April 2008.

 *electronics*

Article

A Hybrid Asymptotic-FVTD Method for the Estimation of the Radar Cross Section of 3D Structures

Alessandro Fedeli *, Matteo Pastorino and Andrea Randazzo

Department of Electrical, Electronic, Telecommunications Engineering and Naval Architecture (DITEN),
University of Genoa, 16145 Genoa, Italy; matteo.pastorino@unige.it (M.P.); andrea.randazzo@unige.it (A.R.)
* Correspondence: alessandro.fedeli@unige.it; Tel.: +39-010-33-52753

Received: 19 October 2019; Accepted: 17 November 2019; Published: 21 November 2019

Abstract: The Finite Volume Time-Domain (FVTD) method is an effective full-wave technique which allows an accurate computation of the electromagnetic field. In order to analyze the scattering effects due to electrically large structures, it can be combined with methods based on high-frequency approximations. This paper proposes a hybrid technique, which combines the FVTD method with an asymptotic solver based on the physical optics (PO) and the equivalent current method (ECM), allowing the solution of electromagnetic problems in the presence of electrically large structures with small details. Preliminary numerical simulations, aimed at computing the radar cross section of perfect electric conducting (PEC) composite objects, are reported in order to evaluate the effectiveness of the proposed method.

Keywords: numerical simulation; finite volume methods; physical optics; radar cross-section

1. Introduction

Several electromagnetic problems require solvers able to characterize complex and multi-scale structures [1–4]. To this end, two main classes of approaches can be identified. The first one is composed by full-wave approaches, which aim to solve Maxwell's equations (or equivalent equations derived from them) without approximations different from the ones introduced by the numerical discretization of the problem. Common solvers belonging to such a class are the method of moments (MoM) [5], the finite-difference time-domain (FDTD) [6] and finite-difference frequency-domain (FDFD) [7] methods, the finite integration technique (FIT) [8], the finite-element method (FEM) [9], and the method of lines (MOL) [10]. Although being very effective in several practical applications, these approaches have the drawback that the computational requirements usually increase significantly when very large (in terms of wavelengths) radiating or scattering structures are considered. In such cases, high-frequency techniques based on asymptotic approximations are frequently adopted. Common methods belonging to this second class are based on the physical optics (PO) approximation and on the geometrical/physical theory of diffraction [9,11].

When targets composed by large and small scatterers are concerned, hybrid methods, in which full-wave and high-frequency techniques are combined together, can also be adopted. In this framework, different hybrid formulations have been proposed in the literature in the past years. For example, the method of moments has been combined with different high-frequency solvers, both for characterizing the radiation of antennas in the presence of large structures [12–15], and for computing the field scattered by large or multi-scale objects [16–18]. Iterative techniques have also been proposed for increasing the accuracy and reducing the computational time [19,20]. Time-domain approaches, such as the FDTD and FIT methods, have been adopted in combination with asymptotic algorithms, too [21–25]. It is worth remarking that, in several cases, one of the main difficulties in developing hybrid methods is represented by the design of a correct interface between the full-wave and asymptotic solver. To partially overcome such a problem, the use of Green's functions, including the scattering contributions from the large scatterers present in the simulated scenario, has also been proposed [26–28].

In this paper, a hybrid approach based on the Finite Volume Time-Domain (FVTD) method [3,29–31], which is a full-wave technique based on a finite-volume discretization of Maxwell's equations, is proposed. The FVTD method has been successfully used in several applications [32–35] concerning both perfect electric conducting (PEC) and dielectric targets, and even in combination with other approaches. For example, full-wave FVTD/FDTD techniques have been proposed in [36,37], whereas in [38], a hybrid FVTD/PO method has been developed and applied to two-dimensional (2D) structures. In the present paper, the FVTD method is combined with an asymptotic technique in order to analyze the radar cross section (RCS) of three-dimensional (3D) structures. In particular, the FVTD solver is used for computing the scattering contributions due to one or more small dielectric or metallic objects, whereas the asymptotic solver is adopted for including the effects of large metallic structures located nearby them. The initial idea concerning such a hybridization has been firstly reported in [39], where it has been used for characterizing antennas near large metallic structures. In that case, however, only sources internal to the FVTD region were handled and only the PO contributions due to the nearby metallic structures were included. In the present work, the proposed hybrid technique is extended for the first time in order to include more scattering contributions from the asymptotic region. To this end, the PO approximation is adopted for computing the scattered fields from smooth surfaces, whereas the Equivalent Current Method (ECM) [11] is used for including the effects of the edges. As far as we know, this is the first time that such a hybrid approach is developed and numerically validated for 3D configurations.

The paper is organized as follows. The formulation of the hybrid method is reported in Section 2. Numerical results aimed at validating the proposed computational technique are provided in Section 3. Finally, conclusions are drawn in Section 4.

2. Formulation of the Hybrid Method

With reference to Figure 1, the computational domain is split into two parts, namely the FVTD region and the asymptotic region. In the first region, which contains an electrically small-size object, the FVTD solver is applied. In the second region, which is used for the electrically large structures, electromagnetic fields are calculated by using the asymptotic technique. In particular, the PO method is applied to compute the scattered field by the surfaces in which the asymptotic region is discretized, whereas edges are taken into account by using ECM.

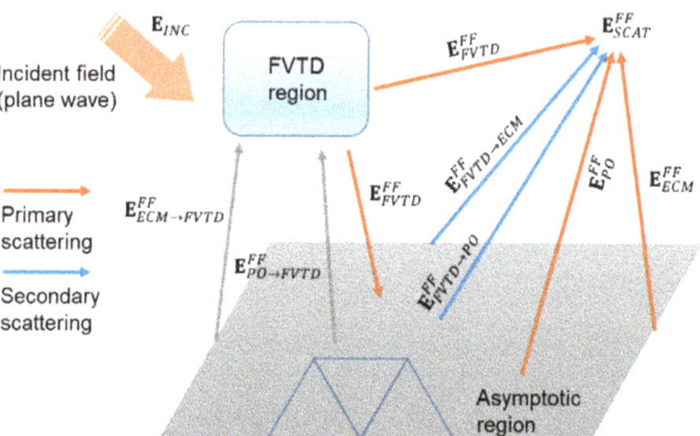

Figure 1. Combined contributions of the two regions.

The computational domain is illuminated by a plane wave, $\mathbf{E}_{INC}(\mathbf{r}) = \mathbf{E}_{INC}^{0} e^{-jk_0 \hat{\mathbf{d}}_t \cdot \mathbf{r}}$, where $\hat{\mathbf{d}}_t$ is the propagation direction, $\mathbf{E}_{INC}^{0}$ describes the electric field amplitude and polarization, and k_0 is the vacuum wavenumber. The overall scattered electric field in the far-field region, indicated as $\mathbf{E}_{SCAT}^{FF}(\mathbf{r})$, is approximately given by the following contributions:

$$\mathbf{E}_{SCAT}^{FF}(\mathbf{r}) \cong \mathbf{E}_{FVTD}^{FF}(\mathbf{r}) + \mathbf{E}_{PO}^{FF}(\mathbf{r}) + \mathbf{E}_{ECM}^{FF}(\mathbf{r}) + \mathbf{E}_{FVTD \to PO}^{FF}(\mathbf{r}) + \mathbf{E}_{FVTD \to ECM}^{FF}(\mathbf{r}) \tag{1}$$

The first term, $\mathbf{E}_{FVTD}^{FF}(\mathbf{r})$, represents the electric field scattered by the object inside the FVTD region, whereas $\mathbf{E}_{PO}^{FF}(\mathbf{r})$ and $\mathbf{E}_{ECM}^{FF}(\mathbf{r})$ denote the primary electric fields scattered by the plane surfaces in which the asymptotic region is discretized and by their edges, respectively. However, these three terms do not take into account the interactions between the two regions. The most significant of these interactions are included in secondary scattered fields, denoted as $\mathbf{E}_{FVTD \to PO}^{FF}(\mathbf{r})$ and $\mathbf{E}_{FVTD \to ECM}^{FF}(\mathbf{r})$. These two terms concern the field scattered by the objects located in the asymptotic region when illuminated by the field produced by the FVTD region.

The contribution produced by the field scattered by the asymptotic region and radiated onto the FVTD region is neglected, assuming that the latter region is significantly larger than the former one. Under this assumption, it is possible to simulate the FVTD region alone, and subsequently combine the result with the contributions due to the asymptotic region. The scattering problem is solved by adopting a proper discretization of both regions, as described in the following sections.

2.1. Scattered Field from the FVTD Region

The primary scattering contribution from the FVTD region is computed using the full-wave FVTD method. We adopt a discretization with tetragonal elementary volumes V_n ($n = 1, \ldots, N$), whose faces are denoted as $S_{n,k}$ ($k = 1, \ldots, 4$) and characterized by outward unit vectors $\hat{\mathbf{n}}_{n,k}$. By defining the vector field $\mathbf{U}(\mathbf{r}, t) = \left(E_x, E_y, E_z, H_x, H_y, H_z \right)^t$ and using the second-order Lax–Wendroff temporal scheme [40], from Maxwell's equations we obtain the following explicit update equations for each nth elementary volume [3],

$$\mathbf{U}_n^{(i+\frac{1}{2})} = \mathbf{U}_n^{(i)} + \frac{\Delta t}{2}\left[\frac{1}{V_n} \sum_{k=1}^{4} \mathbf{\Lambda}_{n,k}^{(i)} S_{n,k} - \mathbf{A}_n^{-1}\mathbf{C}_n\mathbf{U}_n^{(i)} \right] \tag{2}$$

$$\mathbf{U}_n^{(i+1)} = \mathbf{U}_n^{(i)} + \Delta t\left[\frac{1}{V_n} \sum_{k=1}^{4} \mathbf{\Lambda}_{n,k}^{(i+\frac{1}{2})} S_{n,k} - \mathbf{A}_n^{-1}\mathbf{C}_n\mathbf{U}_n^{(i+\frac{1}{2})} \right] \tag{3}$$

where the superscript i represents the time-step index and Δt is its time-width, which is chosen to satisfy the stability criterion [3]. The term $\mathbf{U}_n$ is the field vector computed at the center of the nth tetrahedron, whereas $\mathbf{\Lambda}_{n,k}$ represents the field flux through the surface $S_{n,k}$, which is related to the tangential component of electromagnetic field [1]. By applying the Monotonic Upwind Scheme for Conservative Laws (MUSCL) [1], it is possible to compute the value of the electromagnetic field at the center of each face, which is a quantity required for evaluating $\mathbf{\Lambda}_{n,k}$. The term $\mathbf{A}_n^{-1}\mathbf{C}_n\mathbf{U}_n^{(i)}$ takes into account the presence of media with finite electric conductivities. This term is computed by means of the Additive Induced Source Technique (AIST) [34]. In particular, $\mathbf{A}_n$ is a matrix containing the values of the dielectric permittivity ϵ and magnetic permeability μ inside the nth subdomain, whereas $\mathbf{C}_n$ is a matrix containing the values of the electric conductivity in the same volume element. Detailed definitions of these matrices can be found in [1]. At the boundaries of the simulation domain, Silver–Müller absorbing boundary conditions [1] are applied in order to remove the inward components of the flux at the boundary faces. Near-field to far-field (NFFF) transformations in the time-domain [6] are then used to compute the scattered electric and magnetic fields $\mathbf{E}_{FVTD}^{FF}$ and $\mathbf{H}_{FVTD}^{FF}$ outside the FVTD region, i.e., in test points and in both faces and edges of the asymptotic region. In order to compute the NFFF transformation, a proper Huygens surface is considered inside the FVTD region. Since the subsequent steps of the hybrid procedure are evaluated in the frequency-domain, the Fast Fourier Transform (FFT) is applied to FVTD results in order to compute the scattered field term $\mathbf{E}_{FVTD}^{FF}(\mathbf{r})$ at the considered frequency.

2.2. Scattered Field from the Asymptotic Region

The primary scattered field due to the asymptotic region, which is assumed to be composed of a PEC material, is computed by considering two different contributions. The first one is related to the scattering from surfaces and it is computed by using the PO method [11,41] in the frequency-domain. The scattered field is approximated with the field generated by the current density vector $\mathbf{J}_{PO}$, given by [42]

$$\mathbf{J}_{PO}(\mathbf{r}) = \begin{cases} 2\hat{\mathbf{n}}(\mathbf{r}) \times \mathbf{H}_{INC}(\mathbf{r}) & \text{if } \mathbf{r} \in S_{ill} \\ 0 & \text{otherwise} \end{cases} \tag{4}$$

where $\hat{\mathbf{n}}$ is the outward unit vector perpendicular to the illuminated part of the surface S_{ill} and $\mathbf{H}_{INC}(\mathbf{r}) = \hat{\mathbf{d}}_t \times \eta_0^{-1}\mathbf{E}_{INC}(\mathbf{r})$ is the incident magnetic field vector at a point $\mathbf{r}$ on S_{ill}. Assuming that the illuminated surface S_{ill} is discretized into I triangular faces, S_i, $i = 1, \ldots, I$, the resulting scattered electric field can be expressed as [42]

$$\mathbf{E}_{PO}^{FF}(\mathbf{r}) = \mathcal{L}_{PO}(\mathbf{H}_{INC})(\mathbf{r}) = \frac{\eta_0 e^{-jk_0 r}}{j2\lambda_0 r} \sum_i \left\{ \mathbf{N}_{PO}^i(\hat{\mathbf{r}}) - \left[\mathbf{N}_{PO}^i(\hat{\mathbf{r}}) \cdot \hat{\mathbf{r}}\right]\hat{\mathbf{r}} \right\} \tag{5}$$

where λ_0 is the wavelength, η_0 is the intrinsic impedance of the vacuum, and $\mathbf{N}_{PO}^i$ is the radiation vector related to the ith face, which is given by $\mathbf{N}_{PO}^i(\hat{\mathbf{r}}) = \int_{S_i} \mathbf{J}_{PO}(\mathbf{r}\prime)e^{jk_0\mathbf{r}' \cdot \hat{\mathbf{r}}}d\mathbf{r}'$.

The second contribution is related to the edges. As previously mentioned, the ECM is used [11,41], which assumes that the scattered field due to the edges can be computed in terms of equivalent electric and magnetic line currents. With reference to the above discretization in triangular elements, this part of the scattered field is obtained by summing the contributions of each lth elementary edge [42] between adjacent triangles, i.e.,

$$\mathbf{E}_{ECM}^{FF}(\mathbf{r}) = \mathcal{L}_{ECM}(\mathbf{E}_{INC}, \mathbf{H}_{INC})(\mathbf{r}) = \frac{\eta_0 e^{-jk_0 r}}{j2\lambda_0 r} \sum_l \left\{ \mathbf{N}_{ECM}^l(\hat{\mathbf{r}}) - \left[\mathbf{N}_{ECM}^l(\hat{\mathbf{r}}) \cdot \hat{\mathbf{r}}\right]\hat{\mathbf{r}} - \frac{1}{\eta_0}\hat{\mathbf{r}} \times \mathbf{L}_{ECM}^l(\hat{\mathbf{r}}) \right\} \tag{6}$$

where $\mathbf{N}_{ECM}^l$ and $\mathbf{L}_{ECM}^l$ are the electric and magnetic radiation vectors related to the lth edge, given by $\mathbf{N}_{ECM}^l(\hat{\mathbf{r}}) = \hat{\mathbf{t}}_l \int_{C_l} I_e(\mathbf{r}\prime)e^{jk_0\mathbf{r}' \cdot \hat{\mathbf{r}}}d\mathbf{r}'$ and $\mathbf{L}_{ECM}^l(\hat{\mathbf{r}}) = \hat{\mathbf{t}}_l \int_{C_l} I_m(\mathbf{r}\prime)e^{jk_0\mathbf{r}' \cdot \hat{\mathbf{r}}}d\mathbf{r}'$, respectively, $\hat{\mathbf{t}}_l$ being the unit vector defining the direction of the lth edge C_l. The equivalent electric and magnetic currents I_e and I_m depend upon $\mathbf{E}_{INC}$ and $\mathbf{H}_{INC}$, respectively, and are defined as in [41,43].

2.3. Scattered Fields Due to the Coupling between Regions

The secondary scattered fields $\mathbf{E}_{FVTD \to PO}^{FF}(\mathbf{r})$ and $\mathbf{E}_{FVTD \to ECM}^{FF}(\mathbf{r})$ are due to the coupling between the FVTD and the asymptotic regions. Such contributions are again computed by using the PO approximation and the ECM. However, in this case, the incident field is represented by the primary field scattered by the FVTD region in the points belonging to the asymptotic region. In this paper, we assume that the distance between the asymptotic and the FVTD regions is sufficiently large to allow the use of the far-field approximation for the computation of the fields $\mathbf{E}_{FVTD}^{FF}(\mathbf{r})$ and $\mathbf{H}_{FVTD}^{FF}(\mathbf{r})$, $\mathbf{r} \in S_{ill}$, that illuminate the asymptotic region. Consequently, the coupling terms can be compactly expressed as

$$\mathbf{E}_{FVTD \to PO}^{FF}(\mathbf{r}) = \mathcal{L}_{PO}\left(\mathbf{H}_{FVTD}^{FF}\right)(\mathbf{r}) \tag{7}$$

$$\mathbf{E}_{FVTD \to ECM}^{FF}(\mathbf{r}) = \mathcal{L}_{ECM}\left(\mathbf{E}_{FVTD}^{FF}, \mathbf{H}_{FVTD}^{FF}\right)(\mathbf{r}) \tag{8}$$

3. Numerical Results

In this Section, some preliminary numerical results obtained by using the above hybrid method are reported. The first considered configuration is shown in Figure 2. A PEC sphere of diameter

$d_S = 0.318$ m is located over a PEC plate of sides $l_x = l_y = 1$ m. The distance between the center of the sphere and the plate is denoted as D. A standard spherical coordinate system centered on the sphere is assumed (in the following, as usual, θ and ϕ denote the elevation and azimuth angles, as shown in Figure 2). The incident field is a plane wave propagating in the directions $\phi = 0°$, $\theta \in [0°, 90°]$ and polarized along the elevation direction $\hat{\theta}$. In the FVTD solver, a Gaussian pulse with unit amplitude and time-width at half-maximum of 1 ns has been used. The simulation time has been set equal to 29 ns and the time step is $\Delta t = 3.4$ ps. In this case, the FVTD region is a spherical volume of diameter $d_{FVTD} = 1$ m, which has been discretized into $N_{FVTD} = 90,375$ tetrahedra. The Huygens surface S_H used for computing the NFFF transformations is a sphere of diameter $d_H = 0.6$ m discretized with $N_H = 3572$ triangles. Both the FVTD region and the Huygens surface are centered at the origin of the reference system. The asymptotic region is represented by the plate, which has been discretized with a mesh composed of $N_{AS} = 28$ triangles.

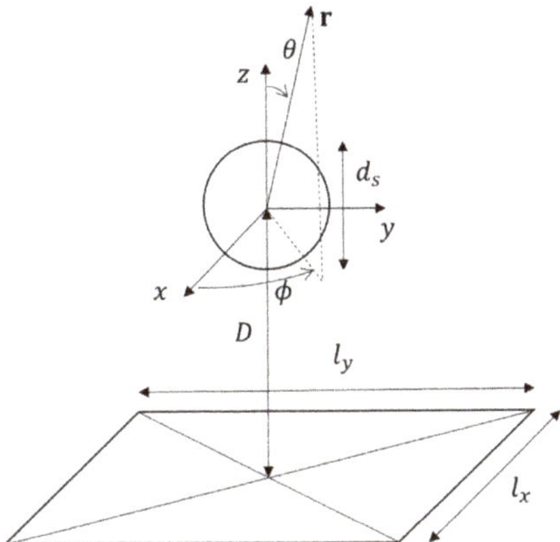

Figure 2. Schematic representation of the first configuration. Sphere over a perfect electric conducting (PEC) plate.

Figure 3 shows the monostatic RCS computed at 750 MHz for $0° \le \theta \le 90°$, and $\phi = 0°$. In this case, $D = 5$ m. In particular, the figure reports the values of the RCS for the two targets (sphere and plate) computed independently (i.e., without mutual interactions) and by exploiting the proposed hybrid approach. In this case, the effects of the mutual interactions between the sphere and the plate, correctly estimated with the hybrid method, are more visible for high elevation angles. It is worth remarking that, even in such a simple configuration, the proposed hybrid technique allows a considerable saving in the computational requirements. Indeed, if the FVTD method were used to simulate the whole scenario, a simulation region containing both the sphere and the plate (with some additional space between targets and boundaries) would be required. For example, assuming a spherical region of side 7 m discretized with a uniform mesh with elements of side length equal to $\lambda/10$ (referred to the maximum frequency of the excitation signal), more than 100 million tetrahedra would be needed, resulting in unfeasible memory requirements on a standard personal computer. Instead, in the considered hybrid technique, only 1.66 GB and 529 MB of RAM are needed for the FVTD and PO/ECM solvers, respectively.

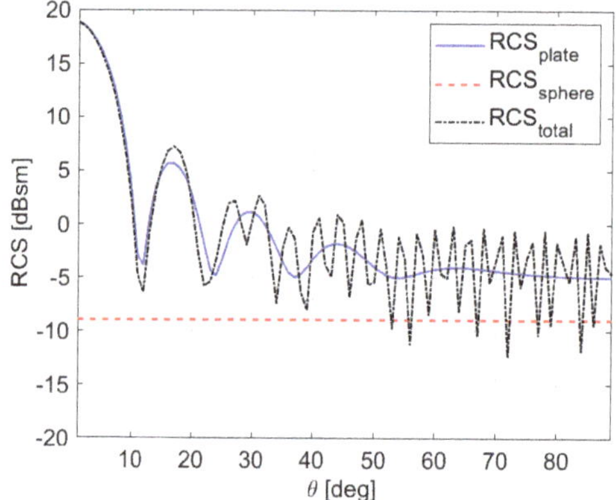

Figure 3. Radar cross section (RCS) of the sphere over plate target. Plate of sides $l_x = l_y = 1$ m located at $D = 5$ m from the sphere. Comparison with the RCS of the sphere and plate alone.

Moreover, in order to validate the use of the NFFF transformations for computing the secondary scattering contributions, the distance between the sphere (i.e., the FVTD region) and the plate (i.e., the asymptotic region) has been varied in the range $D \in [1, 5]$ m. The obtained results have been compared with those provided by the FEKO software (Altair Engineering Inc.) [44]. In particular, the hybrid solver based on the MoM and uniform theory of diffraction (UTD) has been applied considering a mesh of 708 triangular elements. The normalized root mean square error (NRMSE) between the RCS simulated by the proposed hybrid method and by the FEKO solver, for different values of the distance between plate and sphere, are reported in Table 1. Moreover, an example of the behavior of the RCS versus the elevation angle for the case $D = 3$ m is also shown in Figure 4. As can be seen, there is a good agreement between the proposed procedure and the results provided by FEKO. As expected, when the distance between FVTD region and asymptotic region is small, higher errors are present, since the far-field assumptions of the NFFF transformations are no longer satisfied.

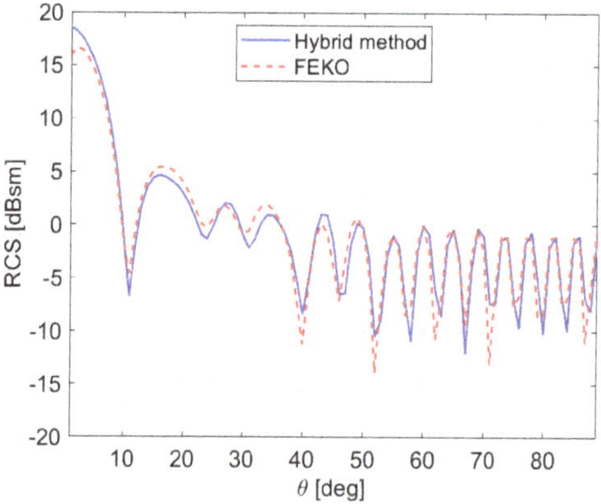

Figure 4. RCS of the sphere over plate target. Plate of sides $l_x = l_y = 1$ m located at $D = 3$ m from the sphere. Comparison with the results provided by the FEKO software.

Table 1. Normalized root mean square error (NRMSE) versus the distance between asymptotic and Finite Volume Time-Domain (FVTD) regions.

Distance [m]	1	2	3	4	5
NRMSE	0.118	0.072	0.055	0.056	0.056

As a further test, the sides of the plate have been varied between 0.5 and 2 m, with $D = 4$ m. In this case, the plate is discretized with $N_{AS} = 76$ triangles. The errors on the computed RCS with respect to the results provided by FEKO are reported in Table 2. As expected, the error is lower when the plate is larger, mainly because the PO method is less effective in simulating the scattered field by small electrical objects.

Table 2. Normalized root mean square error (NRMSE) versus the size of the plate.

Plate side [m]	0.5	1	2
NRMSE	0.142	0.056	0.048

In the second considered test case, a PEC sphere of diameter $d_S = 0.318$ m is located over a square frustum of height $h = 2$ m (Figure 5). The upper base has size $b_x = b_y = 1$ m, and the lower base has side lengths $l_x = l_y = 1.5$ m. The distance between the center of the sphere and the upper base of the frustum is equal to $D = 3$ m. The FVTD region is equal to the one considered in the previous case. The asymptotic region corresponds to the outer surface of the frustum, and it has been discretized with $N_{AS} = 158$ triangles. Figure 6 shows the computed RCS at 750 MHz. The results given by the FEKO full MoM solver (with a mesh composed of 1360 triangular elements) are provided for comparison purposes, as well as FEKO results with hybrid MoM/PO solver with full ray tracing. As can be seen, the agreement between the proposed hybrid method and the full MoM solution, which has been taken as a reference, is quite good (the NRMSE is equal to 0.080), although for some directions the RCS is slightly underestimated. Such differences may be ascribed to the contributions of the field scattered by the asymptotic region and radiated onto the FVTD, which are currently neglected in the developed approach. It is also worth noting that, at least in this case, the obtained results are better than the ones of the FEKO MoM/PO approach, for which the NRMSE with respect to the full MoM is equal to 0.136.

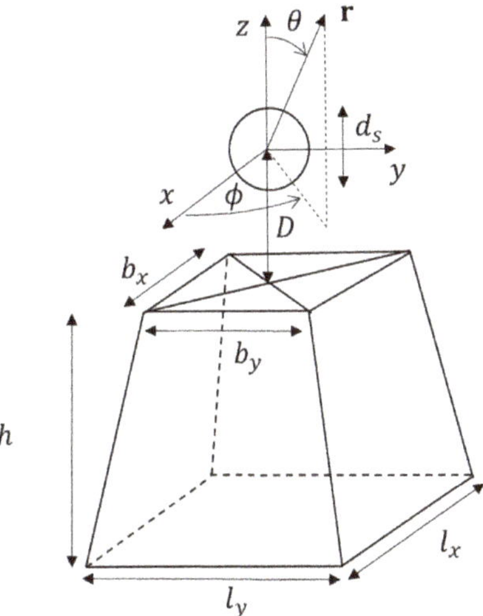

Figure 5. Schematic representation of the second configuration. Sphere over a PEC square frustum.

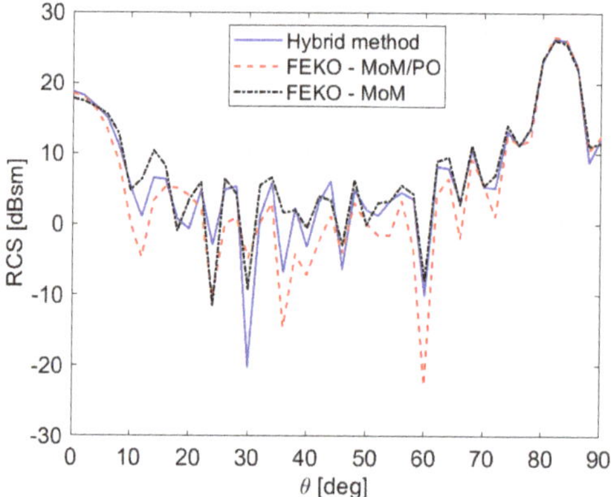

Figure 6. RCS of the sphere over frustum target. Comparison with the results provided by the FEKO software (full method of moments (MoM) and MoM/physical optics (PO) approaches).

4. Conclusions

In this paper, a hybrid approach based on the FVTD method and PO/ECM asymptotic techniques has been presented, with the aim of computing the scattered electromagnetic field by multi-scale objects. The computational domain is split into two parts: An FVTD region and an asymptotic region, which contain small-size and large structures, respectively. The mutual interactions between regions, which produce secondary scattered field contributions, are approximated by considering the most significant terms and far-field interactions. Preliminary numerical simulations, in which the proposed technique is compared with independent electromagnetic simulators for estimating the RCS of simple composite PEC targets, show the effectiveness of the approach. Future developments will be aimed at modeling near-field interactions between the two regions and at validating the technique in the presence of more complex structures and dielectric targets, for which the FVTD method is proven to be particularly effective. The combination of FVTD and asymptotic methods in the time-domain will be considered, too. Such an extension would allow simulations to take into account the contributions due to the field scattered by the asymptotic region and radiated back onto the FVTD region. Moreover, further work will also be devoted to the introduction of multiple reflections through the integration of geometrical optics techniques in order to increase the accuracy of the solution. Finally, the possibility of exploiting proper Green's functions into the FVTD method for including the contributions of large structures, as well as the use of iterative schemes, will be explored.

Author Contributions: Conceptualization, A.F., M.P., and A.R.; methodology, A.F., M.P., and A.R.; formal analysis, A.F., M.P., and A.R.; investigation, A.F., M.P., and A.R.; validation, A.F., M.P., and A.R.

Funding: This research received no external funding.

Acknowledgments: The Authors are grateful to Riccardo Ferri for his support in the preparation of the numerical simulations reported in this paper.

Conflicts of Interest: The authors declare no conflict of interest.

References

1. Rao, S.M. *Time Domain Electromagnetics*; Academic Press: San Diego, CA, USA, 1999; ISBN 978-0-08-051924-1.
2. Davidson, D.B. *Computational Electromagnetics for Rf and Microwave Engineering*, 2nd ed.; Cambridge University Press: Cambridge, UK, 2010; ISBN 978-0-511-91890-2.
3. Fumeaux, C.; Baumann, D.; Leuchtmann, P.; Vahldieck, R. A generalized local time-step scheme for efficient FVTD simulations in strongly inhomogeneous meshes. *IEEE Trans. Microw. Theory Tech.* **2004**, *52*, 1067–1076. [CrossRef]

4. Frezza, F.; Pajewski, L.; Ponti, C.; Schettini, G. Through-wall electromagnetic scattering by N conducting cylinders. *J. Opt. Soc. Am. A Opt. Image Sci. Vis.* **2013**, *30*, 1632–1639. [CrossRef] [PubMed]

5. Harrington, R.F. *Field Computation by Moment Methods*; IEEE: Piscataway, NJ, USA, 1993; ISBN 978-0-7803-1014-8.

6. Taflove, A.; Hagness, S.C. *Computational Electrodynamics: The Finite-Difference Time-Domain Method*, 3rd ed.; Artech House: Boston, MA, USA, 2005; ISBN 1-58053-832-0.

7. Ramahi, O.M.; Subramanian, V.; Archambeault, B. A simple finite-difference frequency-domain (FDFD) algorithm for analysis of switching noise in printed circuit boards and packages. *IEEE Trans. Adv. Packag.* **2003**, *26*, 191–198. [CrossRef]

8. Clemens, M.; Weiland, T. Discrete electromagnetism with the finite integration technique. *Prog. Electromagn. Res.* **2001**, *32*, 65–87. [CrossRef]

9. Paknys, R. *Applied Frequency-Domain Electromagnetics*; John Wiley & Sons: Hoboken, NJ, USA, 2016; ISBN 978-1-118-94054-9.

10. Pregla, R.; Helfert, S. *Analysis of Electromagnetic Fields and Waves: The Method of Lines*; John Wiley & Sons: Hoboken, NJ, USA, 2008; ISBN 978-0-470-05850-3.

11. Knott, E.F.; Shaeffer, J.; Tuley, M. *Radar Cross Section*, 2nd ed.; SciTech Publishing: Raleigh, NC, USA, 2004; ISBN 978-1-891121-25-8.

12. Liu, Z.L.; Wang, X.; Wang, C.F. Installed performance modeling of complex antenna array mounted on extremely large-scale platform using fast MoM-PO hybrid framework. *IEEE Trans. Antennas Propag.* **2014**, *62*, 3852–3858. [CrossRef]

13. Moneum, M.A.A.; Shen, Z.; Volakis, J.L.; Graham, O. Hybrid PO-MoM analysis of large axi-symmetric radomes. *IEEE Trans. Antennas Propag.* **2001**, *49*, 1657–1666. [CrossRef]

14. Obelleiro, F.; Taboada, J.M.; Rodríguez, J.L.; Rubiños, J.O.; Arias, A.M. Hybrid moment-method physical-optics formulation for modeling the electromagnetic behavior of on-board antennas. *Microw. Opt. Technol. Lett.* **2000**, *27*, 88–93. [CrossRef]

15. Çivi, Ö.A.; Pathak, P.H.; Chou, H.T.; Nepa, P. Hybrid uniform geometrical theory of diffraction—Moment method for efficient analysis of electromagnetic radiation/scattering from large finite planar arrays. *Radio Sci.* **2000**, *35*, 607–620. [CrossRef]

16. Braaten, B.D.; Nelson, R.M.; Mohammed, M.A. Electric field integral equations for electromagnetic scattering problems with electrically small and electrically large regions. *IEEE Trans. Antennas Propag.* **2008**, *56*, 142–150. [CrossRef]

17. Vipiana, F.; Francavilla, M.A.; Vecchi, G. EFIE modeling of high-definition multiscale structures. *IEEE Trans. Antennas Propag.* **2010**, *58*, 2362–2374. [CrossRef]

18. Mei, X.; Zhang, Y.; Lin, H. A new efficient hybrid SBR/MoM technique for scattering analysis of complex large structures. In Proceedings of the 2015 IEEE International Conference on Computational Electromagnetics, Hong Kong, China, 2–5 February 2015; pp. 306–308.

19. Chen, M.; Zhao, X.W.; Zhang, Y.; Liang, C.H. Analysis of antenna around nurbs surface with iterative MoM-PO technique. *J. Electromagn. Waves Appl.* **2006**, *20*, 1667–1680. [CrossRef]

20. Liu, Z.L.; Wang, C.F. Efficient iterative method of moments-physical optics hybrid technique for electrically large objects. *IEEE Trans. Antennas Propag.* **2012**, *60*, 3520–3525. [CrossRef]

21. Yang, L.X.; Ge, D.B.; Wei, B. FDTD/TDPO hybrid approach for analysis of the EM scattering of combinative objects. *Prog. Electromagn. Res.* **2007**, *76*, 275–284. [CrossRef]

22. Faghihi, F.; Heydari, H. Time domain physical optics for the higher-order FDTD modeling in electromagnetic scattering from 3-D complex and combined multiple materials objects. *Prog. Electromagn. Res.* **2009**, *95*, 87–102. [CrossRef]

23. Chia, T.T.; Burkholder, R.J.; Lee, R. The application of FDTD in hybrid methods for cavity scattering analysis. *IEEE Trans. Antennas Propag.* **1995**, *43*, 1082–1090. [CrossRef]

24. Lepvrier, B.L.; Loison, R.; Gillard, R.; Pouliguen, P.; Potier, P.; Patier, L. A new hybrid method for the analysis of surrounded antennas mounted on large platforms. *IEEE Trans. Antennas Propag.* **2014**, *62*, 2388–2397.

25. Chou, H.T.; Hsu, H.T. Hybridization of simulation codes based on numerical high and low frequency techniques for the efficient antenna design in the presence of electrically large and complex structures. *Prog. Electromagn. Res.* **2008**, *78*, 173–187. [CrossRef]

26. Burkholder, R.J.; Kim, Y.; Pathak, P.H.; Lee, J.F. Green's function approach for interfacing UTD with FEM for a conformal array antenna on a large platform. In Proceedings of the Fourth European Conference on Antennas and Propagation, Barcelona, Spain, 12–16 April 2010; pp. 1–4.

27. Burkholder, R.J.; Pathak, P.H.; Sertel, K.; Marhefka, R.J.; Volakis, J.L.; Kindt, R.W. A hybrid framework for antenna/platform analysis. *ACES J.* **2006**, *21*, 177–195.

28. Tzoulis, A.; Eibert, T.F. A hybrid FEBI-MLFMM-UTD method for numerical solutions of electromagnetic problems including arbitrarily shaped and electrically large objects. *IEEE Trans. Antennas Propag.* **2005**, *53*, 3358–3366. [CrossRef]

29. Yee, K.S.; Chen, J.S. The finite-difference time-domain (FDTD) and the finite-volume time-domain (FVTD) methods in solving Maxwell's equations. *IEEE Trans. Antennas Propag.* **1997**, *45*, 354–363. [CrossRef]

30. Bonnet, P.; Ferrieres, X.; Issac, F.; Paladian, F.; Grando, J.; Alliot, J.C.; Fontaine, J. Numerical Modeling of Scattering Problems Using a Time Domain Finite Volume Method. *J. Electromagn. Waves Appl.* **1997**, *11*, 1165–1189. [CrossRef]

31. Baumann, D.; Fumeaux, C.; Vahldieck, R. Field-based scattering-matrix extraction scheme for the FVTD method exploiting a flux-splitting algorithm. *IEEE Trans. Microw. Theory Tech.* **2005**, *53*, 3595–3605. [CrossRef]

32. Fumeaux, C.; Baumann, D.; Vahldieck, R. Finite-volume time-domain analysis of a cavity-backed Archimedean spiral antenna. *IEEE Trans. Antennas Propag.* **2006**, *54*, 844–851. [CrossRef]

33. Baumann, D.; Fumeaux, C.; Hafner, C.; Li, E.P. A modular implementation of dispersive materials for time-domain simulations with application to gold nanospheres at optical frequencies. *Opt. Express* **2009**, *17*, 15186–15200. [CrossRef] [PubMed]

34. Bozza, G.; Caviglia, D.D.; Ghelardoni, L.; Pastorino, M. Cell-centered finite-volume time-domain method for conducting media. *IEEE Microw. Wirel. Compon. Lett.* **2010**, *20*, 477–479. [CrossRef]

35. Almpanis, G.; Fumeaux, C.; Vahldieck, R. The trapezoidal dielectric resonator antenna. *IEEE Trans. Antennas Propag.* **2008**, *56*, 2810–2816. [CrossRef]

36. Yee, K.S.; Chen, J.S. Conformal hybrid finite difference time domain and finite volume time domain. *IEEE Trans. Antennas Propag.* **1994**, *42*, 1450–1455. [CrossRef]

37. Ferrieres, X.; Parmantier, J.P.; Bertuol, S.; Ruddle, A.R. Application of a hybrid finite difference/finite volume method to solve an automotive EMC problem. *IEEE Trans. Electromagn. Compat.* **2004**, *46*, 624–634. [CrossRef]

38. Chatterjee, A.; Myong, R.S. Efficient implementation of higher-order finite volume time-domain method for electrically large scatterers. *Prog. Electromagn. Res.* **2009**, *17*, 233–254. [CrossRef]

39. Cerruti, M.; Pastorino, M.; Randazzo, A. A hybrid FVTD-PO approach for the characterization of antennas on large platforms in naval applications. In Proceedings of the 2015 MTS IEEE OCEANS, Genoa, Italy, 8–21 May 2015; pp. 1–5.

40. Lax, P.; Wendroff, B. Systems of conservation laws. *Commun. Pure Appl. Math.* **1960**, *13*, 217–237. [CrossRef]

41. Domingo, M.; Torres, R.P.; Catedra, M.F. Calculation of the RCS from the interaction of edges and facets. *IEEE Trans. Antennas Propag.* **1994**, *42*, 885–888. [CrossRef]

42. Balanis, C.A. *Advanced Engineering Electromagnetics*, 2nd ed.; John Wiley & Sons: Hoboken, NJ, USA, 2012; ISBN 978-0-470-58948-9.

43. Ufimtsev, P.Y. *Fundamentals of the Physical Theory of Diffraction*; IEEE: Hoboken, NJ, USA, 2007; ISBN 978-0-470-09771-7.

44. Electromagnetic Simulation Software|Altair FEKO. Available online: https://altairhyperworks.com/product/FEKO (accessed on 17 November 2019).

 *electronics*

Article

Double-Layer Microstrip Band Stop Filters Etching Periodic Ring Electromagnetic Band Gap Structures

Xuemei Zheng [1,2], Tao Jiang [1], Hao Lu [3] and Yanyan Wang [3,*]

1   College of Information and Communication Engineering, Harbin Engineering University,
    Harbin 150001 China; zhengxuemei@hrbeu.edu.cn (X.Z.); jiangtao@hrbeu.edu.cn (T.J.)
2   School of Electrical Engineering, Northeast Electric Power University, Jilin 132012, China
3   Science and Technology on Complex System Control and Intelligent Agent Cooperation Laboratory,
    Beijing 100074, China; luhao@hrbeu.edu.cn
*   Correspondence: wangyanyan@hrbeu.edu.cn; Tel.: +86-451-8251-9808

Received: 31 March 2020; Accepted: 17 July 2020; Published: 28 July 2020

Abstract: The electromagnetic band gap structure (EBGs) is widely used in microwave engineering, such as amplifiers, waveguides, microstrip filters, due to the fact of its excellent band stop characteristics. In this paper, three kinds of microstrip band stop filters were proposed which were etched with a hexagonal ring EBGs, octagonal ring EBGs and elliptical ring EBGs. Firstly, the etching coefficient of a band stop filter is proposed, and the performance of filters with different etching coefficient was analyzed. Secondly, the equivalent circuit of an EBGs band stop filter is proposed. By comparing the simulation results using advanced design system (ADS) and high frequency structure simulator (HFSS), it was found that the simulation results had the same -10 dB stopband width which verifies the correctness of the equivalent circuit model. Finally, three kinds of microstrip stopband filters were fabricated and measured. The experimental results of the -10 dB stopband width and resonant frequency were in good agreement with the simulation results. The -10 dB stopband fractional bandwidth of the three kinds of microstrip stopband filters was more than 63%. The proposed microstrip band stop filters can be widely used in microwave devices with a wide stopband.

Keywords: electromagnetic band gap structure; microstrip band stop filter; band stop characteristics

---

1. Introduction

The photonic band gap (PBG) structure has a certain periodic structure which can prevent the propagation of microwave in a certain frequency range [1]. The electromagnetic band gap structure (EBGs) originates from the PBG structure which has the characteristics of band resistance and slow wave [2]. In recent years, EBGs has been widely used in microwave device design. In the application of microwave power amplifiers, it has usually been used to improve the efficiency and output power of power amplifiers. In the design of microstrip antennas, it is mainly used to improve the performance of the antenna [3]. In the application of microstrip filters, its superior broad stopband characteristic is suitable for microwave band stop filters [4–9].

Microstrip filters based on a variety of EBG structures are realized by a few techniques, and they mainly include the following categories. The first is to realize the filter with notch by coupling an EBGs, adjusting the coupling distance and the size of the unit EBGs. In Reference [4], a microstrip filter based on the Archimedes spiral EBGs is proposed. In Reference [5], a microstrip filter is realized via a coupling T-shaped resonator, and an asymmetric coupling line band-pass filter is proposed in Reference [6], and the performance of the filter is adjusted by adding coupling variables. In Reference [7], an ultra-wideband bandpass filter based on a coupled EBGs is proposed. The second is based on the periodic EBGs which reduces the passband ripple, increases the stopband width, and achieves better band stop characteristics. For example, Reference [8] proposed a DP–EBG structure to achieve band stop characteristics; Reference [9] proposed a filter embedded in a multi-mode resonator; Reference [10] proposed a filter based on a tapered Cauchy microstrip Koch fractal EBGs; and Reference [11] proposed a filter based on a multi-cycle conical etching EBGs. The third is based on the integration of EBGs and a multi-layer PCB. For example, multi-layer PCB adopts the locally embedded planar EBGs [12] and biplane EBG microstrip filter [13]. The fourth is to optimize the EBGs based on high performance optimization algorithm, and analyze its band resistance characteristics to obtain better band resistance. In Reference [14], Particle swarm optimization (PSO) is used to optimize EBGs, and in Reference [15] PSO is used to optimize the EBG common mode filter by using an artificial neural network.

In this paper, three kinds of double-layer microstrip stopband filters with an etched EBG ring structure are proposed. First, a band stop filter based on a gradient line is proposed. In the upper layer of the dielectric plate, the basic band stop filter was realized by using the butterfly gradient microstrip line. The periodic elliptical ring, hexagonal ring, and octagonal ring were etched on the floor, further increasing the stopband width of −10 dB and the maximum attenuation. Second, the influence of the etching factor of the three band stop filters on the band stop width and attenuation was analyzed, and the equivalent circuit of the band top filter was further analyzed. The correctness of the proposed equivalent circuit model was verified by comparing the simulation and measured results. Finally, the three band stop filters were compared with those proposed in other studies. The proposed filters etched with three kinds of EBGs rings can be applied to the broadband band stop RF devices.

2. Design of Microstrip Band Stop Filter

2.1. Design Principle of Microstrip Band Stop Filter

The three band stop filters designed in this paper were composed of a two-layer EBGs. The top layer was the patch microstrip line of butterfly element, and the bottom layer was etched with three different periodic ring EBGs on the floor. According to Reference [14], when the period of the butterfly element is six, the band stop filter can obtain good stopband characteristics and small passband ripple. The dielectric material was Rogers ro4003, the thickness of the dielectric plate was 1.5 mm, and the relative dielectric constant was 3.55. According to Bragg reflection conditions, Formulas (1)–(4) can be derived. The distance between two adjacent circles on the ground plate was d_1, β was the phase constant of the dielectric plate, λ_g was the wavelength of the waveguide, f_0 was the center frequency of the band resistance, ε_{eff} was the effective dielectric constant plate of the medium, and C was the speed of light. The central frequency of the three filters was 5.2 GHz, the period d_1 and d_2 of the upper butterfly structure were 17.1 mm, and the corresponding w_1 was 3.5 mm and w_2 was 0.2 mm.

$$\beta \times d_1 = \pi \tag{1}$$

$$\beta = \frac{2\pi}{\lambda_g} \tag{2}$$

$$\lambda_g = \frac{C}{f_0 \times \sqrt{\varepsilon_{eff}}} \tag{3}$$

$$d_1 = \frac{\lambda_g}{2} \tag{4}$$

The top layer of the three double-layer band stop filters proposed in this paper adopted the butterfly gradient structure with a thickness of 0.035 mm. The top view and the 3D view are shown in Figure 1a,b, respectively. In order to analyze the effect of w_1 and w_2 on the band stop, HFSS was adopted to sweep the frequency from 2 GHz to 8 GHz for w_1 and w_2.

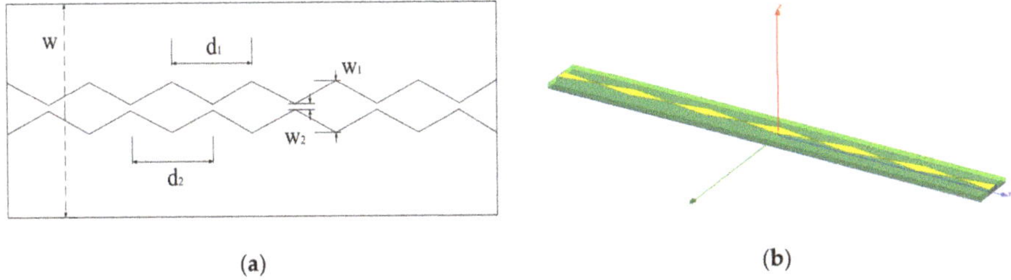

(a)   **(b)**

Figure 1. (**a**) Perspective picture of the microstrip band stop filter based on asymptote; (**b**) simulation results of the microstrip band stop filter based on asymptote.

As shown in Figure 2a, with the increase of w_1, the −10 dB bandwidth increased from 1.59 GHz to 1.91 GHz, the maximum stopband attenuation increased from 19.41 dB to 23.95 dB, the low-frequency ripple decreased from 2.73 dB to 2.29 dB, and the high-frequency ripple increased from 1.99 dB to 3.93 dB. As shown in Figure 2b, with the increase of w_2, the −10 dB bandwidth decreased from 2.01 GHz to 1.49 GHz, the maximum stopband attenuation decreased from 25.39 dB to 18.3 dB, and the low-frequency ripple and high-frequency ripple decreased accordingly.

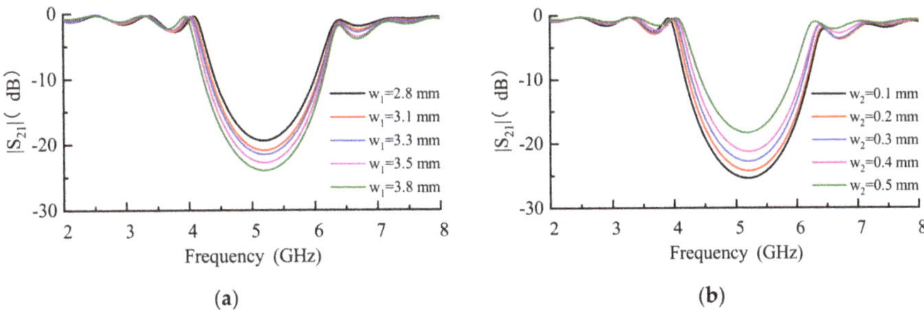

(a)   **(b)**

Figure 2. (**a**) S_{21} with different w_1 by high frequency structure simulator (HFSS); (**b**) S_{21} with different w_2 by using HFSS.

As shown in Figure 3a, with the increase of d_1, the resonance frequency decreased continuously, but the −10 dB bandwidth and the maximum stopband attenuation increased. The simulation of the band stop filter based on the asymptote structure are shown in Figure 3b. The frequency of stopband resonance center was set at 5.2 GHz, and the size of gradient line optimized by HFSS sweeping was as follows: $d_1 = d_2 = 17.1$ mm; $w = 10$ mm; $w_1 = 3.5$ mm; $w_2 = 0.2$ mm; dielectric constant = 3.55; and thickness = 1.5 mm.

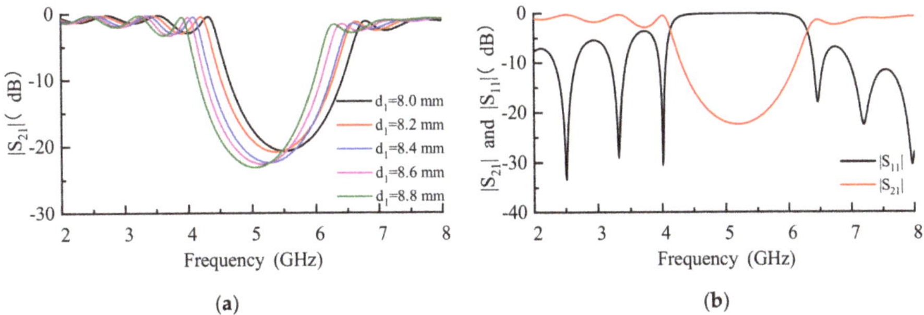

(a)   **(b)**

Figure 3. (**a**) S_{21} with different d_1 by HFSS; (**b**) S_{21} and S_{11} of band stop filter based on a gradient line.

2.2. Band Stop Filter with Etched Periodic Hexagon Ring EBG Structure

The upper layer of the microstrip band stop filter is a butterfly-shaped gradient line which is made of copper with a thickness of 0.034 mm. The floor of the filter is covered with copper and etched with a hexagon ring EBGs. The etched depth is equal to the thickness of copper, which is 0.034 mm. The side length of the outer hexagon is c_1, and the side length of the inner hexagon is c_2. The etching factor of hexagon is $k_h = c_2/c_1$. A perspective view of a microstrip band stop filter etched with a periodic hexagon ring on the floor is shown in Figure 4a, and Figure 4b is a 3D view of a microstrip band stop filter etched with a periodic hexagon ring modeled using HFSS.

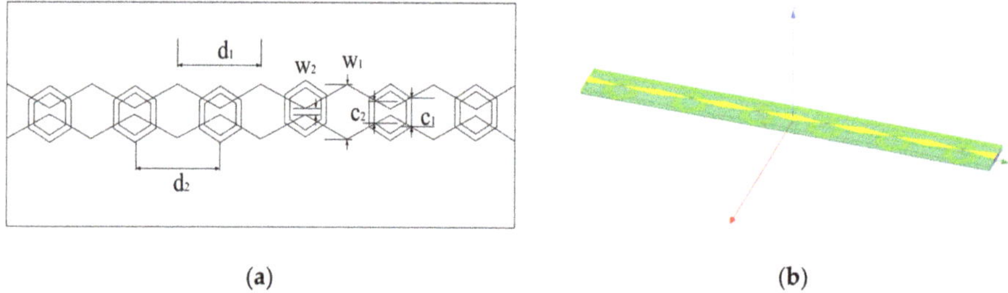

(a)   (b)

Figure 4. (a) Perspective view of microstrip band stop filter etching of a periodic hexagon ring electromagnetic band gap structure (EBGs); (b) 3D picture of a microstrip band stop filter etching of a periodic hexagon ring EBGs.

2.3. Band Stop Filter of Etched Periodic Octagonal Ring EBG Structure

The upper layer of the microstrip band stop filter is a butterfly-shaped gradient line which is made of copper with a thickness of 0.034 mm. The floor of the filter is covered with copper and etched with an octagonal ring EBGs. The etched depth is equal to the thickness of copper which is 0.034 mm. Figure 5a shows a perspective view of a microstrip band stop filter etched with a periodic octagon ring on the floor. The center distance of the outer octagon is e_1, the center distance of the inner octagon is e_2, and the relationship between $k_o = e_2/e_1$. Figure 5b is a 3D diagram of a microstrip band stop filter with periodic octagonal ring etched by HFSS.

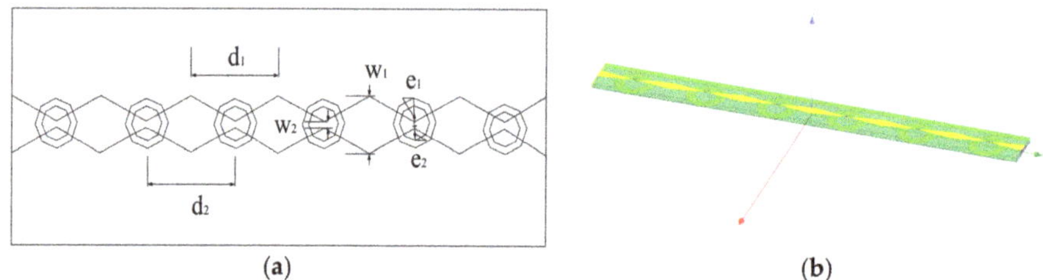

(a)   (b)

Figure 5. (a) Perspective view of a microstrip band stop filter etching of a periodic octagon ring EBGs; (b) 3D picture of a microstrip band stop filter etching of a periodic octagon ring EBGs.

2.4. Band Stop Filter of Etched Periodic Elliptic Ring EBG Structure

The upper layer of the microstrip band stop filter is a butterfly-shaped gradient line which is made of copper with thickness of 0.034 mm. The floor of the filter is covered with copper and etched with an elliptical ring EBGs. The major axis of the outer ellipse ring is a_1, the minor axis is b_1, the major axis of the inner ellipse ring is a_2, and the minor axis is b_2. The ratio of the long axis to the short axis of the outer elliptic ring is r_{a1}, $r_{a1} = b_1/a_1$. The ratio of the major axis to the minor axis is r_{a2}, $r_{a2} = b_2/a_2$. The proportion of the long axis and the short axis of the inner and outer elliptic rings is equal, that is r_a

$= r_{a1} = r_{a2}$. Let the etching factor of elliptical ring be k_e, $k_e = a_2/a_1$. Figure 6a is a perspective view of a microstrip band stop filter etched with a periodic elliptical ring, and Figure 6b is a 3D view of a microstrip band stop filter etched with a periodic elliptical ring modeled with HFSS.

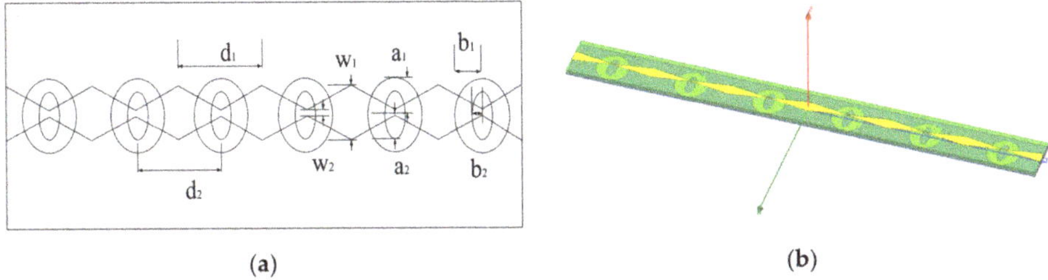

(a)

(b)

Figure 6. (**a**) Perspective view of a microstrip band stop filter etching of a periodic elliptical ring EBGs; (**b**) 3D picture of a microstrip band stop filter etching of a periodic elliptical ring EBGs.

3. Discussions and Results

3.1. Analysis of Band Stop Filter with an Etched Periodic Hexagon Ring EBG Structure

In order to analyze the influence of the etching coefficient of a band stop filter with a hexagon ring, we used HFSS to analyze the frequency sweep of k_h, $c_1 = 3.6$ mm, where the k_h value was 0.5~0.9, and the step size was 0.1. The center frequency was 5.2 GHz, and the optimized parameters were $c_1 = 4.2$ mm and $k_h = 0.22$. As shown in Figure 7a,b, when $c_1 = 3.6$ mm, as k_h increased from 0.5 to 0.9, the maximum attenuation of the stopband decreased from 41.9 dB to 36 dB, the bandwidth of -10 dB decreased from 3.05 GHz to 2.66 GHz, and the resonance frequency of band stop filter decreased from 5.24 GHz to 5.2 GHz.

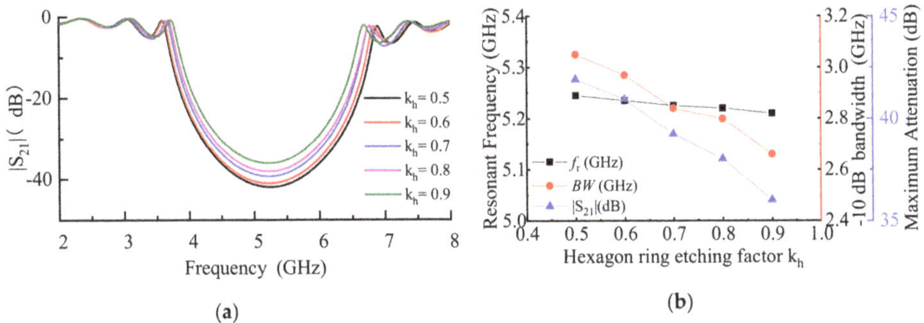

(a)

(b)

Figure 7. (**a**) Simulation with different hexagon ring etching factors k_h using HFSS; (**b**) maximum attenuation and -10 dB fractional bandwidth with different hexagon ring etching factors k_h.

As shown in Figure 8a, when $k_h = 0.3$ and c_1 increased from 3.0 mm to 3.8 mm, the maximum attenuation of band resistance increased from 35.56 dB to 42.87 dB, and the -10 dB bandwidth increased from 2.66 GHz to 3.07 GHz, but the ripple of high and low frequency increased slightly. As shown in Figure 8b, the simulation results of the single-layer graded linear EBGs band stop filter, the filter in Reference [14] and the proposed double-layer band stop filter etching hexagon ring EBGs were compared. The results showed that the performance of the double-layer band-stop filter was obviously better than the former two. When the resonant frequency was 5.2 GHz, $c_1 = 4.2$ mm, $k_h = 0.22$, the -10 dB band stop width was 3.34 GHz and the maximum band resistance attenuation was 48.84 dB.

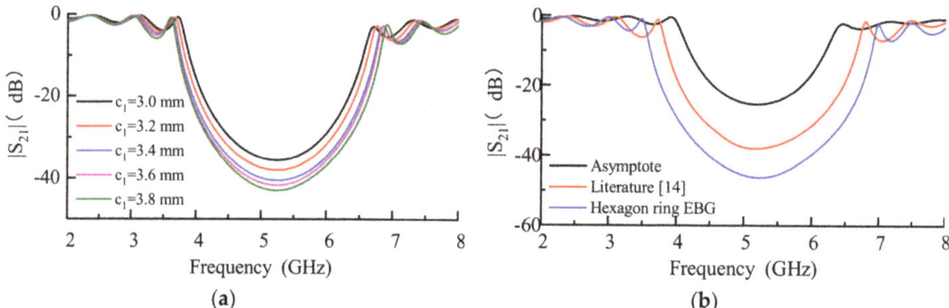

Figure 8. (**a**) Simulation with different c_1 using HFSS; (**b**) comparison of the filter in Reference [14] with the filter etched with a hexagon ring.

Pictures of the top layer and bottom layer of the microstrip band stop filter based on the gradient line structure and etched hexagon ring EBGs are shown in Figure 9a. The comparison between the measured and the simulation is shown in Figure 9b. The results showed that the resonance frequency of the band stop filter was 5.2 GHz, the stopband width of −10 dB was 3.29 GHz, and the maximum attenuation of the stopband was 44.83 dB. The experimental results of the −10 dB stopband width and resonant frequency were in good agreement with the simulation results. The physical test photos and the physical measurement results of the band stop filter etched periodic hexagon ring EBGs are shown in Figure 10a,b. The performance of the band stop filter was tested with the S5180 vector network analyzer by Copper Mountain Technologies. The vector network analyzer was directly connected to the amplitude and phase stabilization test module of N small a type (N-SMA) using the Gore test cable. The calibration was carried out at the SMA port of the test cable of the vector net, and the reference end face was automatically moved to the calibration end face after calibration.

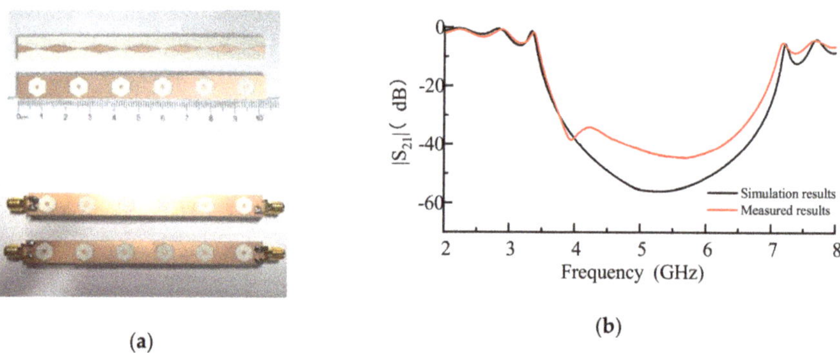

Figure 9. (**a**) Physical photographs of the microstrip band stop filter etched with periodic hexagon ring EBGs welding with an SMA joint; (**b**) comparison of the measured and simulated results of the filter etched with periodic hexagon ring EBGs.

Taking the microstrip band stop filter with etched periodic hexagon ring as an example, the equivalent circuit of band stop filter was analyzed. The equivalent lumped circuit of the microstrip band stop filter based on gradient line EBGs is shown in Figure 11a. The equivalent lumped circuit of the microstrip band stop filter etched with periodic hexagon ring is shown in Figure 11b. The ring EBGs is etched on the floor of the microstrip band stop filter of the asymptote, which is equivalent to adding inductance capacitance (LC) series resonance in the asymptote equivalent circuit and improving the stopband width of the circuit. The three resonant frequencies of the stopband are determined by inductance capacitance series resonance as shown in Formula (5).

$$f_i = \frac{1}{2\pi \sqrt{L_i C_i}} (i = 1, 2, 3) \tag{5}$$

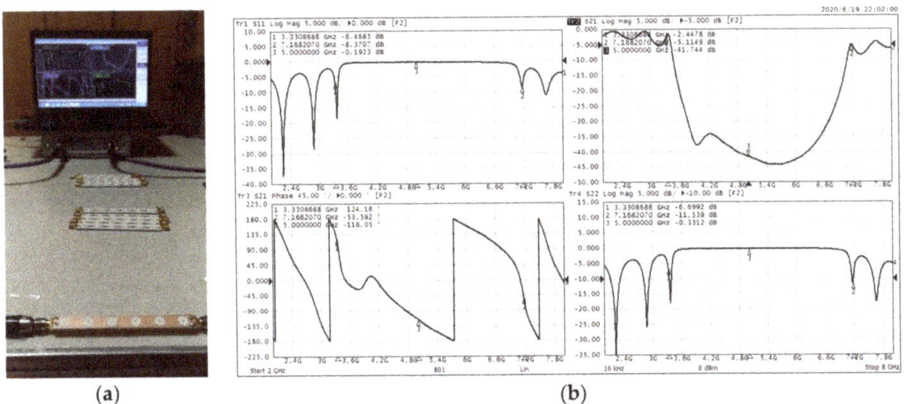

(a)   **(b)**

Figure 10. (a) Physical test photos of the band stop filter etched periodic hexagon ring EBGs; (b) physical measurement results of the band stop filter etched periodic hexagon ring EBGs.

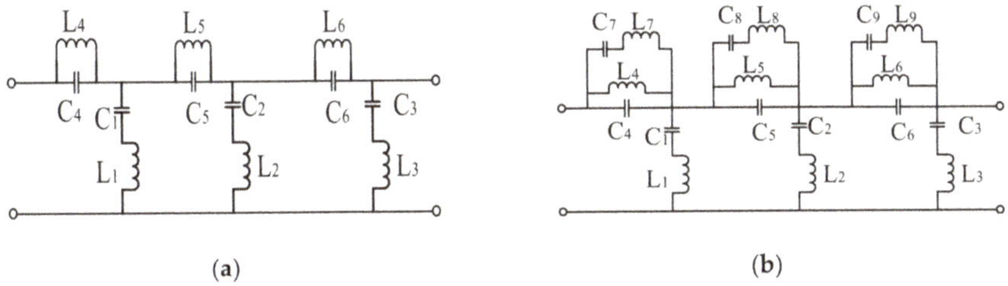

(a)   **(b)**

Figure 11. (a) Equivalent lumped circuit of the microstrip band stop filter based on gradient line EBGs; (b) equivalent lumped circuit of the microstrip band stop filter etched with periodic hexagon ring EBGs.

Advanced design system (ADS) is used to simulate the equivalent lumped circuit of the microstrip band stop filter etched with periodic hexagon ring. The LC parameters are optimized as shown in Table 1. As shown in Figure 12, the simulation results of ads and HFSS show that they have the same -10 dB stopband width, which verifies the correctness of the proposed equivalent circuit.

Table 1. The extracted parameters in Figure 11b by using advanced design system (ADS).

C (pF)	C (pF)	C (pF)	L (nH)	L (nH)	L (nH)
$C_1 = -0.49$	$C_4 = 2.25$	$C_7 = 0.54$	$L_1 = 5$	$L_4 = 2.27$	$L_7 = 6.44$
$C_2 = 0.54$	$C_5 = 1.06$	$C_8 = 1.61$	$L_2 = 1.11$	$L_5 = 3.06$	$L_8 = 1.29$
$C_3 = 0.83$	$C_6 = 2.29$	$C_9 = 0.235$	$L_3 = 0.73$	$L_6 = 1.29$	$L_9 = 10.3$

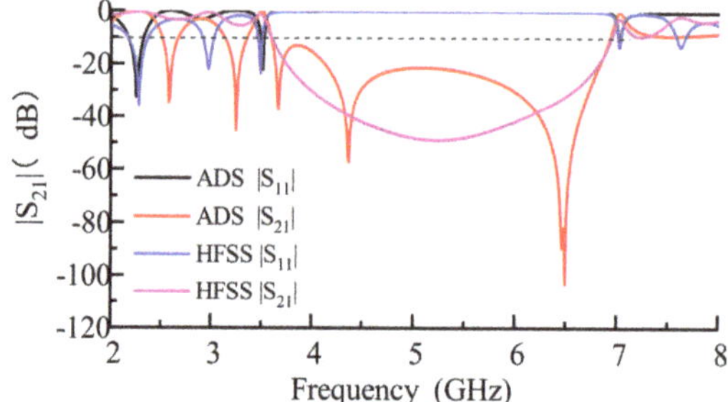

Figure 12. Simulation results of the comparison of the microstrip band stop filter etched with periodic hexagon rings.

3.2. Analysis of Band Stop Filter of Etched Periodic Octagonal Ring EBG Structure

In order to analyze the influence of etching coefficient k_o on the performance of the octagonal band stop filter, the scanning frequency of k_o was analyzed using HFSS, and the sweep range of k_o was from 0.5 to 0.9 in steps 0.1 with the condition $e_1 = 3.6$ mm. The optimized parameters of HFSS were $e_1 = 4.1$ mm and $k_o = 0.25$. As shown in Figure 13a,b, when $e_1 = 3.6$ mm, the maximum attenuation of band stop decreased from 43.1 dB to 36.3 dB, as k_o increased from 0.5 to 0.9, the bandwidth of −10 dB decreased from 3.1 GHz to 2.67 GHz, and the resonance frequency of the band stop filter decreased from 5.24 GHz to 5.2 GHz.

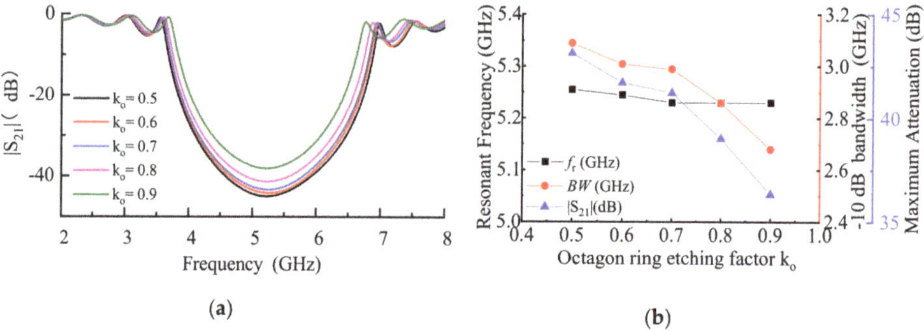

(a)

(b)

Figure 13. (**a**) Simulation with different k_o by using HFSS; (**b**) maximum attenuation and −10 dB fractional bandwidth with different k_o.

As shown in Figure 14a, when $k_o = 0.3$ and e_1 increased from 3.0 mm to 3.8 mm, the maximum attenuation of the band stop increased from 38.1 dB to 44.41 dB, and the −10 dB bandwidth increased from 2.74 GHz to 3.13 GHz, but the high and low frequency ripple slightly increased. As shown in Figure 14b, the HFSS simulation results of single-layer gradual linear EBGs band stop filter, the filter proposed in Reference [14], and the double-layer band stop filter etching octagonal ring EBGs were compared. The results showed that the performance of the double-layer band stop filter was better than the former two filters. When the resonance frequency was 5.2 GHz and $e_1 = 4.1$ mm, $k_o = 0.25$, the band stop width of −10 dB was 3.46 GHz, and the maximum band stop attenuation was 50.25 dB. The simulation results showed that by increasing e_1 or decreasing the etching factor k_o, the band stop width of the filter can be increased and the band stop attenuation can be increased.

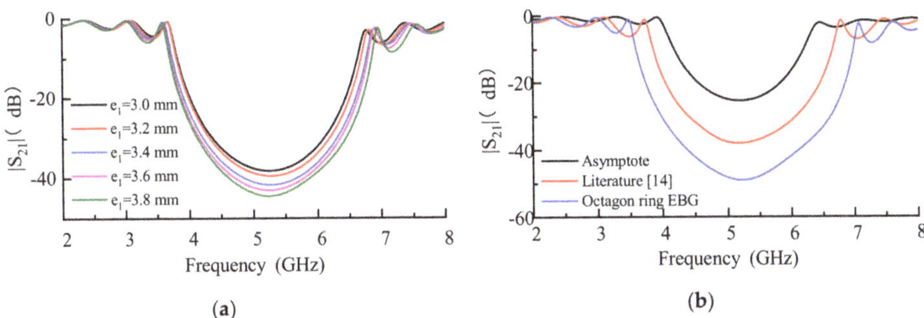

Figure 14. (**a**) Simulation results with different e_1 using HFSS; (**b**) comparison between the filter in Reference [14] and the filter etched with octagon ring.

The pictures of the top layer and bottom layer of the microstrip band stop filter based on the gradient line structure and etched octagonal ring EBGs with SMA joint are shown in Figure 15a. The comparison between the measured and the simulation is shown in Figure 15b. The performance of the band stop filter was tested with the S5180 vector network analyzer by Copper Mountain Technologies.

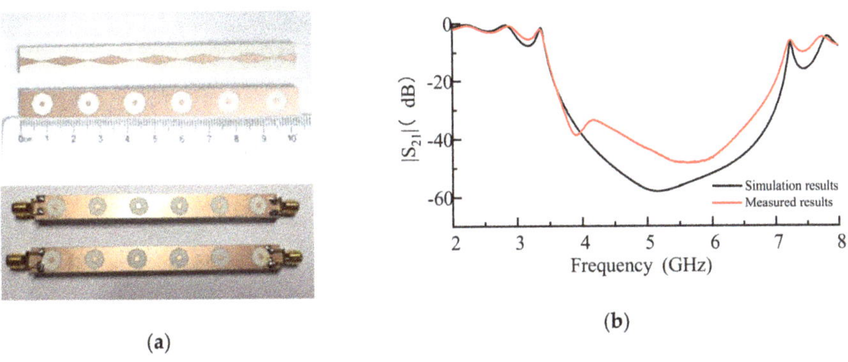

Figure 15. (**a**) Physical photographs of microstrip band stop filter etched periodic octagonal ring EBGs welding with SMA joint; (**b**) comparison of the measured and simulated results of the filter etched periodic octagonal ring EBGs.

The results showed that the resonance frequency of the band stop filter was 5.2 GHz, the stopband width of −10 dB was 3.41 GHz, and the maximum attenuation of the stopband was 49.24 dB. The experimental results of the −10 dB stopband width and resonant frequency were in good agreement with the simulation results. The physical test photos and the physical measurement results of the band stop filter etched periodic hexagon ring EBGs are shown in Figure 16a,b.

3.3. Analysis of Band Stop Filter with Etched Periodic Elliptic Ring EBG Structure

In order to analyze the influence of etching coefficient k_e on the performance of band stop filter etched elliptical ring EBGs. In the case of $r_a = r_{a1} = r_{a2}$, the frequency sweep of k_e was analyzed using HFSS. The relationship between etching coefficient and −10 dB band stop bandwidth, maximum insertion loss, and passband ripple were analyzed.

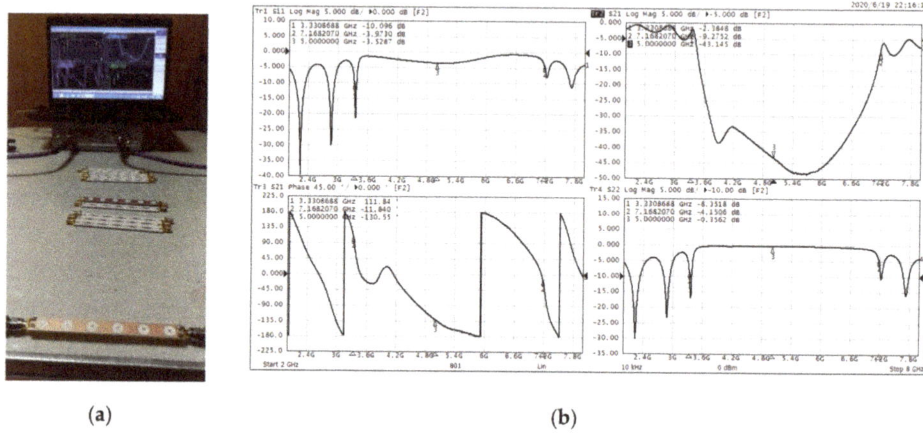

Figure 16. (**a**) Physical test photos of the band stop filter etched periodic octagonal ring EBGs; (**b**) physical measurement results of the band stop filter etched periodic octagonal ring EBGs.

When the resonance frequency was 5.2 GHz, the optimized parameters were $a_1 = 4.1$ mm, $r_a = 0.95$, and $k_e = 0.24$. As shown in Figure 17a,b, $a_1 = 3.6$ mm, $r_a = r_{a1} = r_{a2} = 0.85$, when k_e increased from 0.5 to 0.9, the maximum attenuation of band stop decreased from 42.6 dB to 36.9 dB, the bandwidth of -10 dB decreased from 3.04 GHz to 2.72 GHz, and the resonance frequency of the band stop filter decreased from 5.22 GHz to 5.2 GHz.

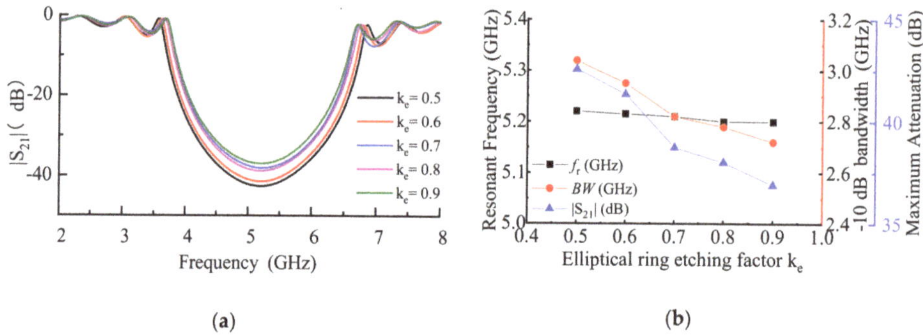

Figure 17. (**a**) Simulation results with different k_e using HFSS; (**b**) maximum attenuation and -10 dB fractional bandwidth with different k_e.

As shown in Figure 18a, as r_a increased from 0.5 to 0.9, the maximum attenuation of the band stop increased from 37.29 dB to 43.25 dB, -10 dB bandwidth increased from 3.77 GHz to 3.69 GHz, low-frequency ripple increased from 3.8 dB to 6.17 dB, and high-frequency ripple increased from 4.46 dB to 9.75 dB.

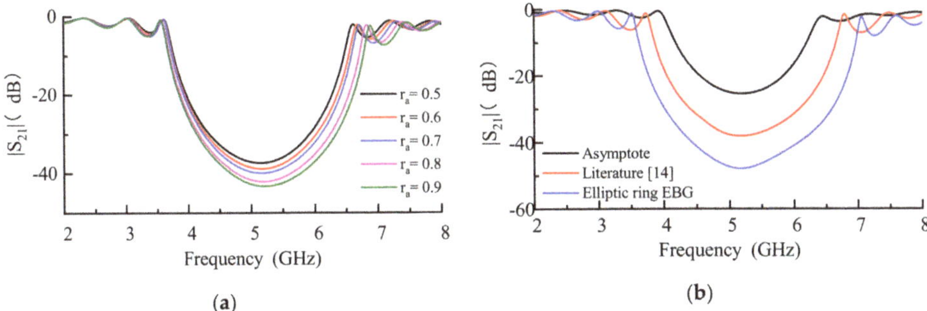

Figure 18. (**a**) Simulation results with different r_a using HFSS; (**b**) comparison of the filter in Reference [14] with the filter etched with the elliptical ring.

As shown in Figure 18b, the microstrip band stop filter based on single-layer gradient EBGs, the band stop filter proposed in Reference [14], and the double-layer band stop filter etched with tthe elliptical ring EBGs were compared. The results showed that the band stop filter etched with elliptical ring EBGs increased the −10 dB stopband width and the maximum attenuation of stopband, but the ripple in the band also increased.

The top layer and bottom layer of the microstrip band stop filter based on the gradient line structure and etched elliptical ring EBGs with SMA joint are shown in Figure 19a. The comparison between the measured and the simulation is shown in Figure 19b. The performance of the band stop filter was tested with the S5180 vector network analyzer by Copper Mountain Technologies. The results showed that the resonance frequency of the band stop filter was 5.2 GHz, the stopband width of −10 dB was 3.46 GHz, and the maximum attenuation of the stopband was 48.57 dB. The experimental results of the −10 dB stopband width and resonant frequency were in good agreement with the simulation results. The physical test photos and the physical measurement results of the band stop filter etched periodic elliptical ring EBGs are shown in Figure 20a,b.

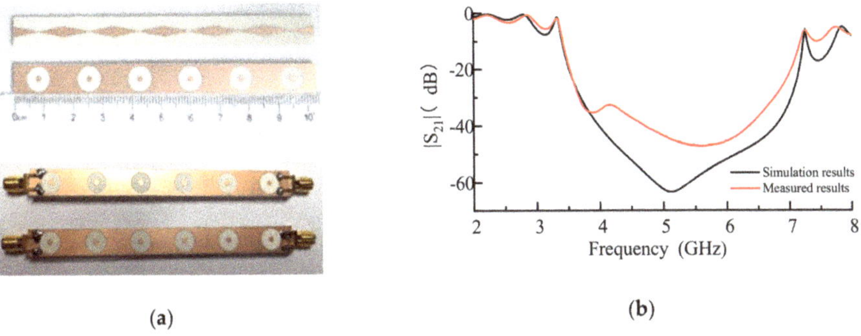

Figure 19. (**a**) Physical photographs of the microstrip band stop filter etched periodic elliptical ring EBGs welding with SMA joint; (**b**) comparison of the measured and simulated results of the filter etched periodic elliptical ring EBGs.

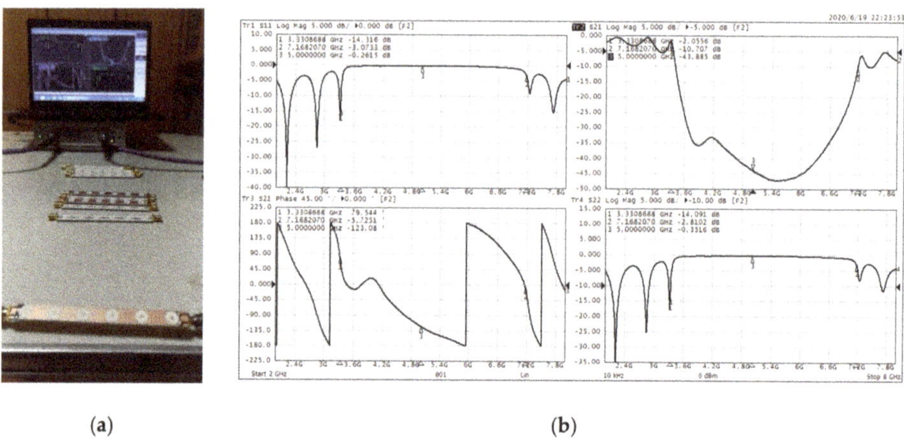

(a)   **(b)**

Figure 20. (**a**) Physical test photos of the band stop filter etched periodic elliptical ring EBGs; (**b**) physical measurement results of the band stop filter etched periodic elliptical ring EBGs.

In order to compare the effect of three etching factors on the stopband characteristics of the three band stop filters, under the condition of $a_1 = c_1 = e_1 = 3.6$ mm, Figure 21a compares the −10 dB bandwidth with the three etching factors, and Figure 21b analyzes the maximum stopband attenuation with the three etching factors. Simulation results show that the curves of the −10 dB bandwidth of stopband and the maximum attenuation of k_e are in good agreement with the curves of k_h which shows that the etching factor k_e and k_h can achieve the same stopband characteristics. Therefore, when $a_1 = c_1$ is determined, the required −10 dB bandwidth and the maximum attenuation of the stopband can be adjusted by changing k_e or k_h.

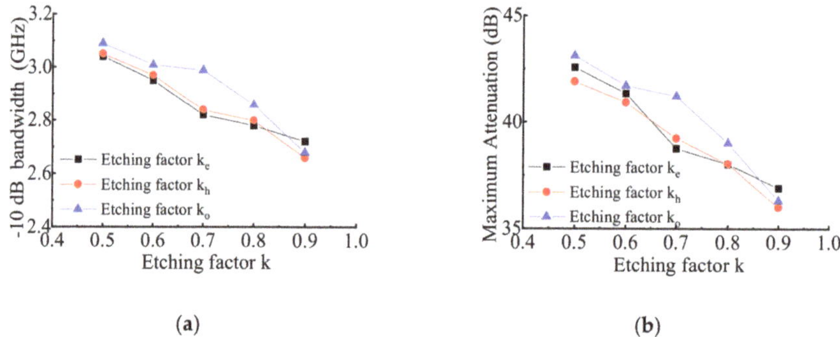

(a)   **(b)**

Figure 21. (**a**) Comparison of simulation of −10 dB bandwidth with three etching factors; (**b**) comparison of the simulations of the maximum attenuation of stopband with three etching factors.

In this paper, three kinds of double-layer microstrip band stop filters were proposed which were designed with a butterfly-shaped gradual microstrip line on the upper layer of the dielectric plate and the EBGs of a periodic elliptic ring, hexagonal ring, and octagonal ring etched on the floor.

4. Conclusions

By etching the EBGs, a wider band stop and a larger band stop attenuation microstrip band stop filter were realized comparing Reference [14]. Given that a_1, c_1, and e_1 were constant, the −10 dB bandwidth and the maximum attenuation of the three microstrip stop band filters decreased with the increase of the etching parameters k_e, k_h, and k_o. The simulation of the designed filter and other band stop filters are compared in Table 2. Compared with References [14] and [15], the stopband center frequency of 5.2 GHz, the −10 dB fractional bandwidths of the three stopband filters proposed in this paper were all greater than 63% which is significantly larger than the former two.

Table 2. Comparison of the proposed filter with other band stop filters.

Reference	Maximum Attenuation of Stopband (dB)	Low Frequency Ripple (dB)	High Frequency Ripple (dB)	Resonant Frequency (GHz)	−10 dB Bandwidths (GHz)	−10 dB Fractional Bandwidths (FBW) %
[14]	42.46	5.05	6.64	5.2	2.73	51%
[16]	41.3	1.2	0.8	5.6	2.55	51.2%
[17]	52	2.3	2.1	4.2	7.8	80.3%
[18]	60.1	6.2	12.3	13	7.5	46.15%
[19]	60.3	4.5	20.6	4.6	4.2	78.1%
[20]	45	31	12.2	6.5	7.8	122.1%
[21]	32.2	5.1	3.7	6.9	1.02	23.3%
[22]	43.1	8.03	7.24	10.2	1.5	14.6%
[23]	48.8	4.13	8.88	10.5	5.03	41%
[24]	44.8	2.3	2.4	30	18	60%
[25]	45.8	1.5	3.6	5.2	4.1	37.7%
[26]	70	5.1	4.81	11.3	17	168.2%
[27]	41.2	6.5	6.8	6.5	3.8	58.4%
[28]	35	2.3	4.3	2.5	1.2	72%
[29]	47.5	1.3	2.5	2.1	1.3	61.5%
[30]	35.6	0.7	0.8	5.2	0.6	11.5%
[31]	70	0.4	0.6	4	1.5	37.5%
Etching elliptic ring EBGs	44.83	4.98	7.34	5.2	3.49	65.78%
Etching hexagon ring EBGs	49.24	5.48	9.45	5.2	3.34	63.31%
Etching octagon ring EBGs	48.57	5.27	7.97	5.2	3.46	65.28%

In this paper, three kinds of double-layer microstrip band stop filters were proposed. The same butterfly gradient microstrip line was used in the upper layer of the dielectric plate, and the periodic elliptical ring, hexagonal ring, and octagonal ring were etched on the floor. Compared with References [14] and [15], the three band stop filters designed in this paper had wider stopband widths and larger maximum attenuations.

In the case of $c_1 = e_1$ and $k_h = k_o$, the trend of maximum attenuation and relative bandwidth of the band stop filter etched with elliptical ring EBGs and the band stop filter etched with hexagon ring EBGs were consistent.

The simulation and measurement results show that the maximum attenuation and the stopband width of −10 dB can be improved by reducing the corrosion factor of the three microstrip band stop filters. The fractional stopband bandwidths of the three microstrip band stop filters were all over 63%, which can be used in radio frequency (RF) devices with wide stopband bandwidth requirements. In practical application, on the basis of ensuring performance, the filter size should be further reduced and the cost should be reduced.

Author Contributions: Conceptualization, methodology, software, validation, formal analysis, investigation, resources, collection and drawing of simulation data and writing—original draft preparation, X.Z.; Writing—review and editing, visualization, supervision, T.J.; Physical testing and testing report, H.L.; Project administration, funding acquisition, Y.W. All authors have read and agreed to the published version of the manuscript.

Funding: This paper was funded by the National Natural Science Foundation of China (61803356), the International Exchange Program of Harbin Engineering University for Innovation-Oriented Talents Cultivation. This work was partially supported by the National Key Research and Development Program of China (2016YFE0111100), Key Research and Development Program of Heilongjiang (GX17A016), the Science and Technology Innovative

Talents Foundation of Harbin (2016RAXXJ044), the Natural Science Foundation of Beijing (4182077), and China Postdoctoral Science Foundation (2017M620918).

Acknowledgments: This paper was funded by the International Exchange Program of Harbin Engineering University for Innovation-Oriented Talents Cultivation. This paper was funded by the National Natural Science Foundation of China (61803356).

Conflicts of Interest: The authors declare no conflict of interest. The founding sponsors had no role in the design of the study; in the collection, analyses, or interpretation of data; in the writing of the manuscript, and in the decision to publish the results.

References

1. Balakin, A.V.; Bushuev, V.A.; Mantsyzov, B.I.; Ozheredov, I.A.; Petrov, E.V.; Shkurinov, A.P.; Masselin, P.; Mouret, G. Enhancement of sum frequency generation near the photonic band gap edge under the quasiphase matching conditions. *Phys. Rev. E* **2001**, *63*, 046609. [CrossRef] [PubMed]
2. Kim, M. A Compact EBG Structure with Wideband Power/Ground Noise Suppression Using Meander-Perforated Plane. *IEEE Trans. Electromagn. Compat.* **2015**, *57*, 595–598. [CrossRef]
3. Hassan, M.A.M.; Kishk, A.A. Bandwidth Study of the Stacked Mushroom EBG Unit Cells. *IEEE Trans. Antennas Propag.* **2017**, *65*, 4357–4362. [CrossRef]
4. Zheng, X.; Jiang, T. Triple Notches Bandstop Microstrip Filter Based on Archimedean Spiral Electromagnetic Bandgap Structure. *Electronics* **2019**, *8*, 964. [CrossRef]
5. Zheng, X.; Pan, Y.; Jiang, T. UWB Bandpass Filter with Dual Notched Bands Using T-Shaped Resonator and L-Shaped Defected Microstrip Structure. *Micromachines* **2018**, *9*, 280. [CrossRef] [PubMed]
6. Parvez, S.; Sakib, N.; Mollah, M. Advanced Investigation on EBG Structures: A Critical Analysis to Optimize the Performance of Asymmetric Couple-Line Bandpass Filter. In Proceedings of the 2015 IEEE International Conference on Electrical, Computer and Communication Technologies, Tamil Nadu, India, 5–7 March 2015; pp. 1–6.
7. Martinez-Iranzo, U.; Moradi, B.; Garcia-Garcia, J. Ultrawideband bandpass filters based on coupled electromagnetic band gap structures. *Microw. Opt. Technol. Lett.* **2015**, *57*, 2857–2859. [CrossRef]
8. Kurra, L.; Abegaonkar, M.P.; Koul, S.K. Equivalent Circuit Model of Resonant-EBG Bandstop Filter. *IETE J. Res.* **2015**, *62*, 17–26. [CrossRef]
9. Wong, S.W.; Zhu, L. EBG-Embedded Multiple-Mode Resonator for UWB Bandpass Filter with Improved Upper-Stopband Performance. *IEEE Microw. Wirel. Components Lett.* **2007**, *17*, 421–423. [CrossRef]
10. Ruiz, J.D.; Martínez, F.L.; Hinojosa, J. A flat-passband and wide-stopband low-pass filter based on tapered Cauchy microstrip Koch fractal EBG structure. In Proceedings of the 2014 8th International Congress on Advanced Electromagnetic Materials in Microwaves and Optics, Lyngby, Denmark, 25–28 August 2014; pp. 139–141.
11. Chen, Y.-C.; Liu, A.-S.; Wu, R.-B. A Wide-stopband Low-pass Filter Design Based on Multi-period Taper-etched EBG Structure. In Proceedings of the 2005 Asia-Pacific Microwave Conference Proceedings, Suzhou, China, 4–7 December 2005; pp. 1–5.
12. Li, L.; Chen, Q.; Yuan, Q.; Sawaya, K. Ultrawideband Suppression of Ground Bounce Noise in Multilayer PCB Using Locally Embedded Planar Electromagnetic Band-Gap Structures. *IEEE Antennas Wirel. Propag. Lett.* **2008**, *8*, 740–743. [CrossRef]
13. Lee, Y.H.; Huang, S.Y. Electromagnetic Compatibility of a Dual-Planar Electromagnetic Band-Gap Microstrip Filter Structure. In Proceedings of the 2006 17th International Zurich Symposium on Electromagnetic Compatibility, Singapore, 27 February–3 March 2006; pp. 561–564.
14. Yu-Bo, T.; Yue, D.; Zhi-Bin, X.; Sha, S.; Tao, P. Frequency characteristics of electromagnetic bandgap structure with bow-tie cells and its optimal design based on particle swarm optimization. *IEEJ Trans. Electr. Electron. Eng.* **2012**, *8*, 63–68. [CrossRef]
15. Orlandi, A. Notice of Retraction: Multiple Objectives Optimization for an EBG Common Mode Filter by Using an Artificial Neural Network. *IEEE Trans. Electromagn. Compat.* **2017**, *60*, 507–512. [CrossRef]
16. Ruiz, J.D.D.; Viviente, F.L.M.; Melcon, A.A.; Hinojosa, J. Substrate Integrated Waveguide (SIW) With Koch Fractal Electromagnetic Bandgap Structures (KFEBG) for Bandpass Filter Design. *IEEE Microw. Wirel. Components Lett.* **2015**, *25*, 160–162. [CrossRef]

17. Ruiz, J.D.D.; Hinojosa, J.; Viviente, F.L.M. Optimisation of chirped and tapered microstrip Koch fractal electromagnetic bandgap structures for improved low-pass filter design. *IET Microw. Antennas Propag.* **2015**, *9*, 889–897. [CrossRef]

18. Joodaki, M.; Rezaee, M. Coplanar Waveguide (CPW) Loaded with an Electromagnetic Bandgap (EBG) Structure: Modeling and Application to Displacement Sensor. *IEEE Sens. J.* **2016**, *16*, 3034–3040. [CrossRef]

19. Panaretos, A.H.; Werner, D.H. A Note on the Isolation Performance of Nonuniform Capacitively Loaded Mushroom-Type EBG Surfaces within a Parallel Plate Waveguide. *IEEE Trans. Antennas Propag.* **2015**, *63*, 5175–5180. [CrossRef]

20. Zhu, H.-R.; Sun, Y.-F.; Wu, X.-L. A Compact Tapered EBG Structure with Sharp Selectivity and Wide Stopband by Using CSRR. *IEEE Microw. Wirel. Components Lett.* **2018**, *28*, 771–773. [CrossRef]

21. Chu, H.; Shi, X. Ultra-wideband bandpass filter with a notch band using EBG array etched ground. *Microw. Opt. Technol. Lett.* **2011**, *53*, 1290–1293. [CrossRef]

22. Simsek, S.; Rezaeieh, S.A. A design method for substrate integrated waveguide electromagnetic bandgap (SIW-EBG) filters. *AEU Int. J. Electron. Commun.* **2013**, *67*, 981–983. [CrossRef]

23. Hassan, S.M.S.; Mollah, N. Identical performance from distinct conventional electromagnetic bandgap structures. *IET Microw. Antennas Propag.* **2016**, *10*, 1251–1258. [CrossRef]

24. Karim, M.; Liu, A.-Q.; Alphones, A.; Zhang, X.; Yu, A. CPW band-stop filter using unloaded and loaded EBG structures. *IEE Proc. Microw. Antennas Propag.* **2005**, *152*, 434–440. [CrossRef]

25. Boutejdar, A.; Elhani, S.; Bennani, S.D. Design of a novel slotted bandpass-bandstop filters using U-resonator and suspended multilayer-technique for L/X-band and Wlan/WiMax applications. In Proceedings of the 2017 International Conference on Electrical and Information Technologies (ICEIT), Rabat, Morocco, 15–18 November 2017; pp. 1–7.

26. Mallahzadeh, A.R.; Rahmati, B.; Alamolhoda, M.; Sharifzadeh, R.; Ghasemi, A.H. Ultra wide stop band LPF with using defected microstrip structures. In Proceedings of the 2012 6th European Conference on Antennas and Propagation (EUCAP), Prague, Czech Republic, 26–30 March 2012; pp. 1–3.

27. Kim, C.; Shrestha, B.; Son, K.C. Wideband bandstop filter using an sir based interdigital capacitor. *Microw. Opt. Technol. Lett.* **2018**, *60*, 2530–2534. [CrossRef]

28. Nasiri, B.; Errkik, A.; Zbitou, J.; Tajmouati, A.; Elabdellaoui, L.; Latrach, M. A New Compact and Wide-band Band-stop Filter Using Rectangular SRR. *TELKOMNIKA Indones. J. Electr. Eng.* **2018**, *16*, 110–117. [CrossRef]

29. Kong, M.; Wu, Y.; Zhuang, Z.; Liu, Y. Narrowband balanced absorptive bandstop filter integrated with wideband bandpass response. *Electron. Lett.* **2018**, *54*, 225–227. [CrossRef]

30. La, D.-S.; Lu, Y.; Sun, S.; Liu, N.; Zhang, J. A novel compact bandstop filter using defected microstrip structure. *Microw. Opt. Technol. Lett.* **2010**, *53*, 433–435. [CrossRef]

31. Zhu, Q.Y.; Meng, X.L. Design and Optimization of Microstrip Stub Bandstop Filter Based on ADS. *Appl. Mech. Mater.* **2015**, *743*, 233–238. [CrossRef]

Article

Electromagnetic Field Levels in Built-up Areas with an Irregular Grid of Buildings: Modeling and Integrated Software

Luca Schirru [1], Filippo Ledda [2], Matteo Bruno Lodi [3,*], Alessandro Fanti [3,*], Katiuscia Mannaro [4], Marco Ortu [3] and Giuseppe Mazzarella [3]

[1] National Institute for Astrophysics (INAF), Cagliari Astronomical Observatory, Via della Scienza 5, 09047 Selargius, Italy; luca.schirru@inaf.it
[2] MetaCell Ltd., LLC, Boston, MA 02142, USA; filippo@metacell.us
[3] Department of Electric and Electronic Engineering, University of Cagliari, 09123 Cagliari, Italy; marco.ortu@unica.it (M.O.); mazzarella@unica.it (G.M.)
[4] Department of Mathematics and Computer Science, University of Cagliari, 09123 Cagliari, Italy; katiuscia.mannaro@unica.it
* Correspondence: matteob.lodi@unica.it (M.B.L.); alessandro.fanti@unica.it (A.F.)

Received: 11 April 2020; Accepted: 5 May 2020; Published: 6 May 2020

Abstract: The knowledge of the electromagnetic field levels generated by radio base stations present in an urban environment is a relevant aspect for propagations and coverage issues, as well as for the compliance to national regulations. Despite the growing interest in the novel fifth generation (5G) technology, several aspects related to the investigation of the urban propagation of the Global System of Mobile Communication (GSM), third generation (3G), and fourth generation (4G) mobile systems in peculiar non-rural environments may be improved. To account for irregular geometries and to deal with the propagation in hilly towns, in this work we present an enhanced version of the COST231-Walfisch–Ikegami model, whose parameters have been modified to evaluate the path loss at distances greater than 20 meters from the radio base station. This work addressed the problem of providing an effective, reliable, and quantitative model for the estimation of electromagnetic field levels in built-up areas. In addition, we also developed and tested a pre-industrial software prototype whose aim is to make the estimated electromagnetic field levels available to the key players in the telecom industry, the local authorities, and the general population. We validated the proposed model with a measurement campaign in the small urban and irregular built-up areas of Dorgali (Nuoro), Cala Gonone (Nuoro), and Lunamatrona (Cagliari) in Sardinia (Italy).

Keywords: radio propagation in an urban environment; electromagnetic field level; narrow band measurements; software prototype

---

1. Introduction

From the early 1980s, worldwide urban environments have witnessed a thorough technological innovation driven by telecommunication systems development. The contemporary users demand a high-quality experience, supplied by a continuous improvement of the services in order to ensure new mobile network functionalities. This results in an overwhelming diffusion of commercial cellular mobile networks, especially those operating in built-up environments in the cellular ultra-high frequency (UHF) bands [1]. Hence, the accurate prediction of the electromagnetic (EM) field levels produced by one or more radio base stations (RBSs) installed in a given urban area, the quantitative assessment of the coverage, and the estimation of loss factors in propagation paths, from the transmitter to the receiver, became determinant aspects for the design of the mobile network or for the evaluation of the EM pollution in any urban environment of interest. Indeed, it is possible to verify if inhabitants are exposed to physical parameters in the limit prescribed by current regulations [2,3]. Moreover, today scientists must communicate scientific evidence clearly. The government agencies must inform people about safety regulations and policy measures, and concerned citizens must decide to what extent they are willing to accept such risk. In this process, it is important that communication among the stakeholders be done clearly and effectively [3]. In order to respond to the needs of mobile network design and the requirements of exposure regulations, theoretical models are able to predict efficiently how EM signal propagates represent valuable and reliable tools for these aims.

It is possible to obtain evaluations of path loss and interference considerations with a suitable propagation model [4,5]. Such a model for propagation in urban environments allows us to avoid long and expansive measuring campaigns, thus constituting a valuable tool for both coverage and exposure assessment. The propagation model must be chosen depending on the peculiar ambient to characterize. Hence, it can be utilized in both the planning and the design phase of the radio systems or when verifying the coverage aims and quality of the service (QoS). To define the model as valid, it is necessary to account for the essential features (e.g., altimetry, characteristics of buildings and streets) of the propagation scenario in an effective and accurate way, i.e., by summarizing the pivotal aspects of specific parameters.

Moreover, the mechanism of propagation for the selected environment has to be correctly described and modeled [4,5]. On the other hand, it is worth noting that the features of the linked antennas deployed in the given scenario must be included in the mathematical description of the types of urban propagation.

In the literature, several propagation models have been proposed. Depending on the descriptive approach of the selected urban environment, the numerical tools for the evaluation of field levels, these models are classified into two types. The former class contains statistical models, which makes use of parameters expressed as average values. The statistical models are used to study wide areas, where the exact evaluation of parameters is expansive. The second model type refers to the so-called deterministic models, which, on the other hand, make use of parameters with exact values, thus being useful for narrow urban areas. From the analysis of the state-of-the-art, it is possible to highlight that several studies were performed with the aim of achieving a correct description of the evaluation of the EM field produced by an RBS together with validation based on an adequate measuring campaign to efficiently monitor the exposure to the EM source. Indeed, a comparison between the EM field values derived from empirical–statistical models, namely, the COST231-Hata model and the COST231–Walfisch–Ikegami (C231WI) model [4,5], and experimental validation through a narrowband measurement campaign was carried out [1]. An interesting and useful study was conducted for the case of the city of Turin (Italy) [6], where the exposure of the population to EM fields generated by RBSs was assessed by comparing measured values with appropriate instrumentations and theoretical values derived from a model which did not consider attenuation and reflection from buildings. Furthermore, a thorough analysis was performed in the city of Monselice (Padova, Italy) with the aim of predicting field levels, considering the alteration due to the presence of buildings (attenuation and reflection), using two different approaches: the former through the expression of the far-field, the latter with the ray-tracing technique [7]. Reference [8] presented work about a comparison of the theoretical evaluation and measured values of power nearby a specific RBS.

The interest of the scientific community is also addressed in the monitoring of EMF levels in urban environments in order to assess the degree of population exposure to UHF EM fields. In the literature, several works focused on the use and description of peculiar and specific measurement systems, and experimental protocols can be found, as shown in Table 1. In Reference [9,10], the authors proposed an innovative monitoring system based on a wireless sensor network (WSN) able to keep under constant control the overall and cumulative EM field in the area of interest. The purpose of the system called SEMONT (Serbian ElectroMagnetic field MOnitoring Network) was the development of a useful tool for national and local agencies of Serbia for environmental protection, especially to keep under control electromagnetic pollution and to asses real-time exposure of the population. In Reference [11,12], the authors presented a study based on the RF radiation levels produced by radio base stations in Serbia and focused the attention on the antennas set at roof height, which are the most common in the analyzed area. These studies [11,12] led to the valuable result of verifying that exposure levels were within limits prescribed by international standards. The problem of field estimation and EM monitoring is a worldwide issue and hence in other countries similar research was performed. For instance, in Turkey the concerns about EM exposure resulted in an indoor and outdoor measuring campaign to assess safety conditions for several inhabitants of a specific area in proximity to a GSM radio base station [13]. In Reference [14], it is possible to observe the results of a study carried out in the rural area of Bari (Italy) focused on the quantification of the effects of exposure to EM, generated by an RBS, on agricultural workers. With reference to the same topic, in [15–17] the results derived from the measurement of power density were compared with EM exposure levels in public areas in Nigeria, in a metropolitan zone in India, and in several areas of China, respectively. The monitoring of electromagnetic field strength from RBSs in urban environments was validated with narrowband measurement campaigns using a spectrum analyzer [17]. As a matter of fact, from this discussion, the experimental monitoring approach is certainly more expansive from an economical and temporal point of view compared with the theoretical and predictive way discussed previously. Therefore, for both the coverage assessment and the exposure concerns, a validated theoretical model constitutes a valuable tool.

Besides the model development and accuracy, as reported in Table 1, in the literature, relevant attention is paid to the translation of the propagation scheme into integrated software for displaying and sharing the field levels in a given area. Indeed, by using optical geometry and the geometric theory of diffraction, hence describing field propagation in terms of beams, a deterministic model was used to design software able to evaluate field levels in urban environments [18]. From the simplified model of antennas transmitting in free space, Windows-based software to predict EM field levels was developed [19].

Among the most relevant models used to predict radio signal propagation in urban environments, the COST 231 Hata model [4,5,20,21] and the COST 231 Walisch-Ikegami model [4,5,22–24] are used for any prediction at distances greater than 20 m from the RBS, as can be seen in Table 1. The COST 231 Hata model, which is an empirical and statistical technique, considers as main parameters the height of the RBS and of the terminal mobile, as well as the distance to the observation point. Therefore, the C231WI model results are more accurate, with respect to other models results. In fact, it can be classified as an empirical–statistical model (see Table 1), which makes use of a better and more careful description of the urban environment, thanks to the different approach to the propagation mechanism by diffraction by roofs and the roof–street linking/coupling [23]. Descriptive parameters for the environment are the average height and distance of separation between adjacent buildings, the width of the street, and the angle between the street and the line of conjunction of the RBS and the mobile station. These are all statistical parameters, and they assume that buildings in an urban center are arranged in a regular grid.

The fundamental working hypotheses of the C231WI can be considered acceptable in medium-sized and large cities; whereas, for the highly recurrent case of irregular, variable, hilly cities and small towns, the C231WI model would surely lose accuracy and effectiveness. The aim of this work is the development, validation, and use of a theoretical model which can describe the propagation of EM fields generated by GSM, 3 G, and 4 G RBSs (at frequencies of 944.2 MHz, 1847.8 MHz, and 2142.4 MHz) in urban scenarios of small, hilly towns with irregular street geometry and small houses with different shapes and heights. In this way, by modifying the definition of the propagation model parameters, it is possible to take into account the plethora of configurations and cases of interest for the coverage estimation and exposure assessment.

This estimation of EM fields from RBSs, using a propagation model based on a modified version of the C231WI, is used to develop an integrated software system with a dedicated mobile application in order to share and visualize the estimated EM fields in a given area, thus informing the telecom companies, the local authorities, and the general population.

Table 1. Summary table of the state-of-the-art of propagation models and field level estimation.

Name	Category	Coverage	Scenario	Country	Year	Ref.
COST231-Hata Mode	Empirical	150 MHz–1.5 GHz d > 20 m	Urban	-	1980	[4,21]
COST231-Walfisch–Ikegami	Empirical Statistical	900 MHz–1.8 GHz d > 20 m	Urban	-	1999	[4,5,23,24]
Prasad et al.	Deterministic Statistical Empirical	1.8 GHz d > 20 m	Urban Suburban	India	2011	[1]
Anglesio et al.	Deterministic	100 kHz–3 GHz d > 20 m	Urban	Italy	2001	[6]
Giliberti et al.	Deterministic (Ray tracing)	3 MHz–3 GHz	Urban Suburban	Italy	2009	[7]
Miclaus and Bechet	Deterministic	900 MHz d > 20 m	Urban	Romania	2007	[8]
SEMONT	Empirical	700 MHz–2.6 GHz	Suburban	Serbia	2014–2020	[9–12]
Çerezci et al.	Empirical	900 MHz–2.1 GHz	Urban	Turkey	2015	[13]
Pascuzzi and Santoro	Deterministic	900 MHz–1.8 GHz d > 20 m	Urban Suburban	Italy	2015	[14]
Ojuh et al.	Deterministic	900 MHz	Rural	Nigeria	2015	[15]
Saravanamuttu et al.	Statistical Empirical	540 kHz–2.6 GHz d > 20 m	Urban	India	2015	[16]
Zheng and Zhigang	Deterministic	30 MHz–3 GHz	Urban	China	2015	[17]

In light of these premises, our work aims to develop a theoretical model for describing the propagation of UHF EM fields produced by one or more radio base stations in specific urban scenarios and at the same time increase awareness of the population at the EM field level in the interested zones. Originally, our research was initiated under an umbrella project called "Onde Chiare" and was supported by the Regione Autonoma Sardegna. This project aimed to answer the need of information about EM fields by the general public while satisfying the requirement of lowering the monitoring efforts of regulatory and local agencies. Accordingly, the first objective of our work was to develop a mathematical framework serving as a design tool for cellular networks and to predict coverage performances. Furthermore, the second objective was the development of a Web-based platform of mobile phone base stations and their emissions and simultaneously of a mobile application linked to this in order to give direct access to data in a user-friendly, reliable, up-to-date, and timely way. Our system aims to provide a common information basis for decision makers and the public and therefore presents the values of EM field levels to the interested population and checks whether the exposure to EMF is likely to be exceeded. In this way, it would be possible to activate procedures to reduce levels when they exceed the attention thresholds.

This paper makes the following contributions to the literature. First, it focuses on relevant models used to predict radio signal propagation in urban environments and then defines an enhanced model for the estimation of loss factors in propagation paths in order to predict accurately the EM field level produced by one or more radio base stations installed in specific urban environments. Secondly, it presents our pre-industrial software prototype called the Onde Chiare System (OCS) oriented to support the knowledge in the electromagnetic field and obtain general information about antennas and regulations. Then, it demonstrates with an experimental validation that the modified version of the COST231-WI model can deal with the built-up scenarios of hilly, largely variable, and small, irregularly arranged towns. The experimental validation is fundamental in order to use the proposed model and test the developed software system. Finally, a discussion of its implications and the conclusions are presented.

2. The Estimation of EM Field Levels

Basically, the field level depends on two factors, namely, the path loss between the RBS and the field point, and the RBS antenna gain in all directions. To calculate the electric field levels in any desired position, the following far-field equation was used [25]

$$E = \sqrt{\frac{8\pi\eta_0 P_t G(\theta,\varphi)}{\lambda^2 L}} \tag{1}$$

where E is the far field in V/m, η_0 is the vacuum impedance, P_t is the transmission power of the RBS transmitter antenna expressed in Watts, $G(\theta,\varphi)$ is the transmitter antenna gain as a function of vertical and horizontal angles in degrees, and L is the attenuation path loss of the electromagnetic signal.

In an urban scenario, there is more than one active RBS. For this reason, we must compute the field amplitude Ep for each RBS using the previous formula and then add the corresponding power density, since the fields of different RBSs are uncorrelated. Therefore, the total EM field is obtained using [25]

$$E_{Tot} = \sqrt{\sum_{p=1}^{n} E_p^2} \tag{2}$$

where n is the number of active RBSs. Regarding the transmitter antenna gain, it must be included in the model. In this work, the term $G(\theta,\varphi)$ was derived with the knowledge of the horizontal and vertical radiation patterns, which are provided by the manufacturers in the datasheets. Typically, the fields for the main E- and H-planes are known. However, the estimation of the electric field should be performed for a given arbitrary point. It is therefore necessary to derive the entire radiation pattern. Among the available methods, in this work, the gain of the transmitting antenna is used as input to the 3D interpolation algorithm from [25,26] in order to derive the full pattern and hence calculate the field levels in a given point. An explicative scheme of the reconstruction procedure is shown in Figure 1 for the case of a Kathrein 742212 antenna.

Modified Version of the COST 231– Walfisch–Ikegami Model

As discussed in the Introduction, a statistical and empirical model is preferable in order to respect and account for the actual topology of the built-up area under analysis in the model. Furthermore, in this work, a model valid for the UHF frequency range and for distances greater than 20 m from the RBS would be considered for coping with hilly and irregular built-up environments. The model which fulfills all these requirements is the C231WI model [4,5,23,24]. Indeed, for the frequency range from 800 MHz to 2000 MHz, for a set of mobile-to-RBS distances from 0.02 km to 5 km, RBS height from 4 m to 50 m, and for mobile height from 1m to 3 m, the COST231WI model allows the path loss evaluation considering the following parameters [4]:

- the height of the buildings in the given scenario (h_{roof});
- the width of the roads in the built-up area (w);
- the building separation (b);
- the road orientation with respect to the radio path (φ).

Therefore, the term L in Equation (1) is a function of these parameters, i.e., $L = L(h_{roof}, w, b, \varphi)$ [4]. A remark is in order. Indeed, in their classical form, the C231WI parameters could be representative of the local field behavior only for a regular urban environment with almost similar buildings having similar features, located in a regular and ordered grid [4]. This is a relevant limitation, considering that the majority of cases, especially in the Italian scenario, are represented by small, hilly towns with irregular arrangements of buildings having significant variability in their height [7,11–19]. When the best, regular configuration is analyzed, the theoretical estimation of the path loss differs by about +3 dB ± 4–8 dB from the measured values, in the case of RBS antennas with heights above the rooftop level [23,24]. Since, in the literature, the analysis of C231WI performances for the cases of non-regular grids, buildings with largely variable height, and set in a non-plane, hilly area has been poorly investigated, to date the error and the deviation of the predicted field levels with respect to measurements are not known. Therefore, employing the C231WI model would lead to unreliable theoretical predictions of EM field values, thus implying a noticeable bias in coverage prediction or in the exposure assessment. Furthermore, for the C231WI model, it is known that the error becomes larger when $h_{base} \approx h_{roof}$, especially with respect to the case when $h_{base} >> h_{roof}$. Moreover, the C231WI has the shortcoming of dealing for distances very close to the source, i.e., the model effectiveness is scarce for $h_{base} << h_{roof}$. Therefore, there is room for improvement.

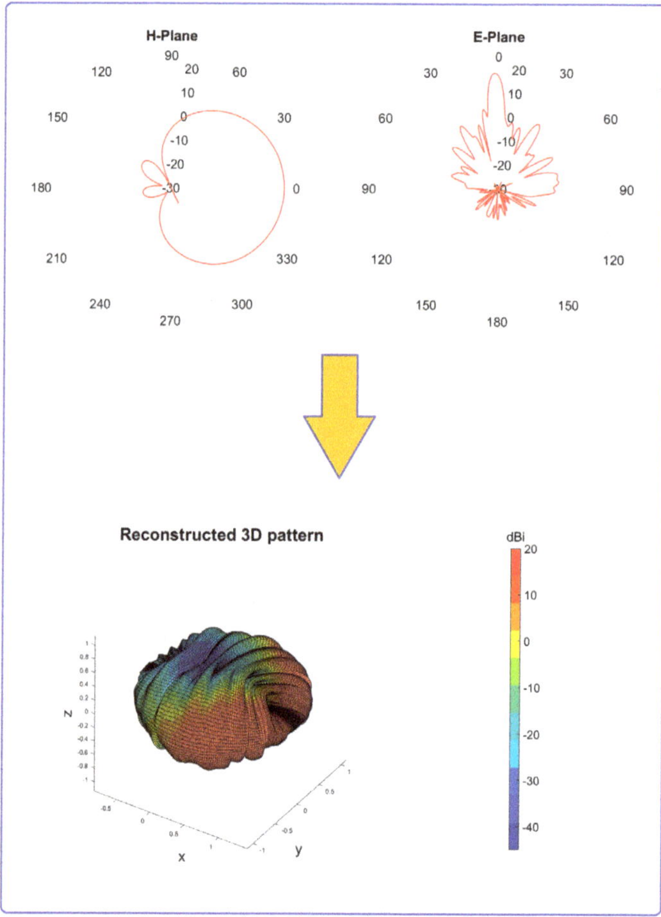

Figure 1. Example of the inputs and output of the algorithm for the 3D reconstruction of the antenna gain [25,26]. The case of the Kathrein 742212 antenna is presented.

It is possible to overcome the aforementioned limitations of the C231WI model by rephrasing and redefining the model parameters, as done in [25]. Instead of using the mean value of the buildings in the grid of interest, h_{roof} should be defined as the mean of the height of the buildings which are crossed by the propagation path, considering the segment which joins the RBS antenna and the ground below the mobile, as shown in Figure 2a [4]. This new definition of the roof height allows us to account for a hilly built-up environment [25], while ensuring the possibility of describing flat cities [5]. This is the built-up environment. Then, parameter w should be assumed to be equal to the width of the street where the receiving mobile is located. A clarification is in order. It is possible to interpret w as the actual road width (w_A) or as the length of the propagation path inside that road (w_p), as shown in Figure 2b [4]. Both possible definitions were tested and verified [25]. Indeed, w can be interpreted as the equivalent ray description of the propagation path (w_p) or as the length of propagation toward the mobile inside a parallel-plane waveguide, with the two buildings as walls, which as the outcome that the dispersion of such a "modal" propagation is an unknown function of w_A [25]. The findings from [25] demonstrated that an actual road with ($w = w_p$) is the modeling strategy for which the prediction error is lower, with respect to the other w parameter evaluations. As regards the term b, in the proposed rephrased model, the arithmetic mean of the separation distances between buildings that are crossed by the beam in its propagation path is a more appropriate definition, as shown in Figure 2c. Finally, φ is redefined as the angle between the propagation path and the last building wall crossed by it before reaching the observation point (see Figure 2c). With these new set of parameters, it is possible to estimate the electric field level in a given urban area using Equation (1).

The modified version of the C231WI model is used to derive the field values in a given urban area. In this study, the model is validated and then used for the irregular and hilly towns of Dorgali (NU), Cala Gonone (NU), and Lunamatrona (CA), Italy. The Onde Chiare software uses the proposed model to derive by request the EM field value at the user location. In this way it is possible to monitor the electromagnetic field levels produced RBSs in urban environments and share this information with the interested stakeholders.

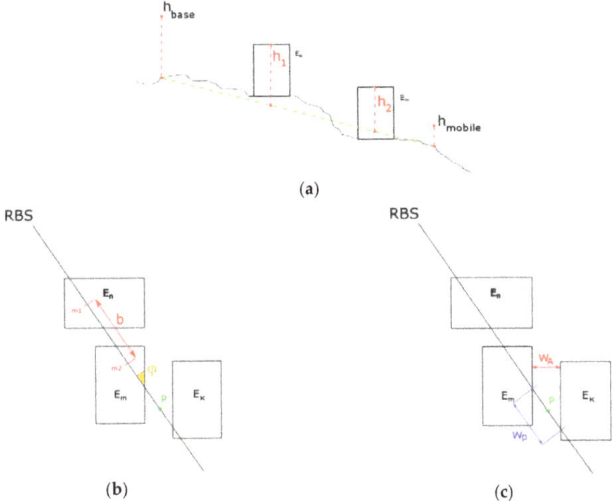

Figure 2. New definition of parameters for the COST231–Walfisch–Ikegami model [24]: (**a**) Schematic drawing of the re-defined parameter hroof to account for the hilliness of a given urban scenario; (**b**) Representation of the two possible definitions of parameter w. In this work, the definition $w = w_p$ is used [24]. (**c**) Definition and representation of the parameters b and φ. The rectangular boxes, named respectively, En, Em, and Ek, are the buildings crossed by the electromagnetic field during its propagation path from the RBS (see label in black) to the point P (see label in green). In addition, h_{base}, h_{mobile}, and $h_{1,2}$ represent the RBS height, the point P height, and the building height, respectively. Finally, the w, b, and φ parameters are the width of the roads in the built-up area, the building separation, and the road orientation with respect to the radio path, respectively

3. Onde Chiare Project

As introduced previously, the "Onde Chiare" project aimed to implement a model for the estimation of loss factors in propagation paths in order to accurately predict EM field levels produced by one or more RBSs installed in particular urban environments. Moreover, it had the goal of developing a regional database of mobile phone base stations and their emissions, while presenting to the public the values of EM field levels in order to show that the exposure to EM fields is below the prescribed limits. In this way it would be possible to activate procedures to reduce levels when they exceed the attention thresholds.

In the context of non-ionizing radiation, data communication as well as its acquisition has a significant importance. The acquired data may help in several functions. They can describe the environment quality and provide information to the local government for urban and territorial planning, but also to the public about the exposure to radio frequency, thus also reducing the concern about upgrades and reconfigurations of mobile phone RBSs. In this direction the project developed a Web-based platform in order to give direct access to data in a user-friendly, reliable, up-to-date, and timely way, providing a common information basis to decision makers and the public.

3.1. System Design

In detail, the software system named Onde Chiare System (OCS) is a pre-industrial prototype setting up a collaborative software platform of services oriented to support the knowledge in the electromagnetic field and obtain general information about antennas and regulations. The general functioning scheme of OCS is shown in Figure 3.

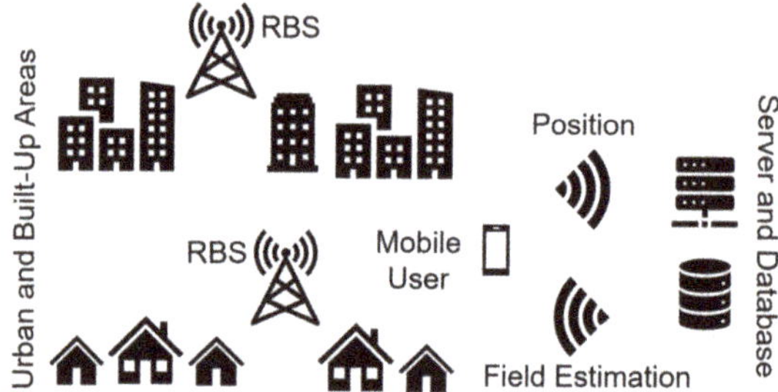

Figure 3. General scheme for the Onde Chiare system. The user can access the database, using an Android device, to check geolocalized information in order to recover the set of surrounding RBSs. The server elaborates the electric field estimation using the modified version of the C231WI model with the geographical and topological data and then sends the output to the user.

OCS has an online register containing the data of the cellular RBS (GSM, 3G, and 4G) installed over the regional territory of Sardinia (Italy). The system aims to support the planning of actions for administrative and regulatory purposes, both for the estimation of the electromagnetic field levels and the evaluation of the population exposure conditions. To address these requirements, the OCS provides a Web application (web app) and a mobile application (mobile app). The Onde Chiare web app maps the geographical distribution of the authorized radio transmitters in Sardinia and provides an information sheet for each of these stations. The map is based on the data supplied by Arpa Sardegna [27]—the regional public service agency responsible for environmental monitoring. This public agency supports the competent authorities in the planning, authorization, and sanctions in the environmental field and it has been tasked with collecting information about health concerns in relation to exposure from EM fields. In particular, the Onde Chiare web app is not only a data register that collects information of mobile phone base stations submitted by the telecommunication operators to the authorities, it also allows the users to observe the results of EM field levels estimated using the model described in the previous section, Section 2. Data are displayed on the map, but they can also be downloaded in various formats for further elaboration. The web app enables the government to manage the territorial and urban transformation processes from the environmental and social perspectives and citizens to gain valuable information about health risks of mobile communications.

OCS provides different access levels: public and private. For privacy and data protection issues, the private problem has different levels of accessibility: for example, restricted areas have been created for the municipalities and the mobile operators with password-controlled access. In practice, the local authority in its area of expertise can view the full specifications of the plants in its territory and can have only public data, not belonging to the municipality itself. An example of the information available using the Onde Chiare system is shown in Figure 4. The database is based on a geographic information system (GIS) giving the antenna mast positions on a normal geographic map. To access the data, the user can type a specific address or the city, and once this is entered, the database will display the map showing an icon marker for each cellular base station in that area. Users can click on the icon and some information will be displayed, as shown in Figure 4b. The Onde Chiare system displays a map that includes both the location of the mobile antennas with technical information and the measurement performed to assess levels of exposure in the surrounding of the masts. Further information on mobile communications technologies and regulations, potential health risks, and research investigations about exposure to radiation are available in .pdf format. The data register can be visualized on mobile devices via a mobile app called the Onde Chiare app [28].

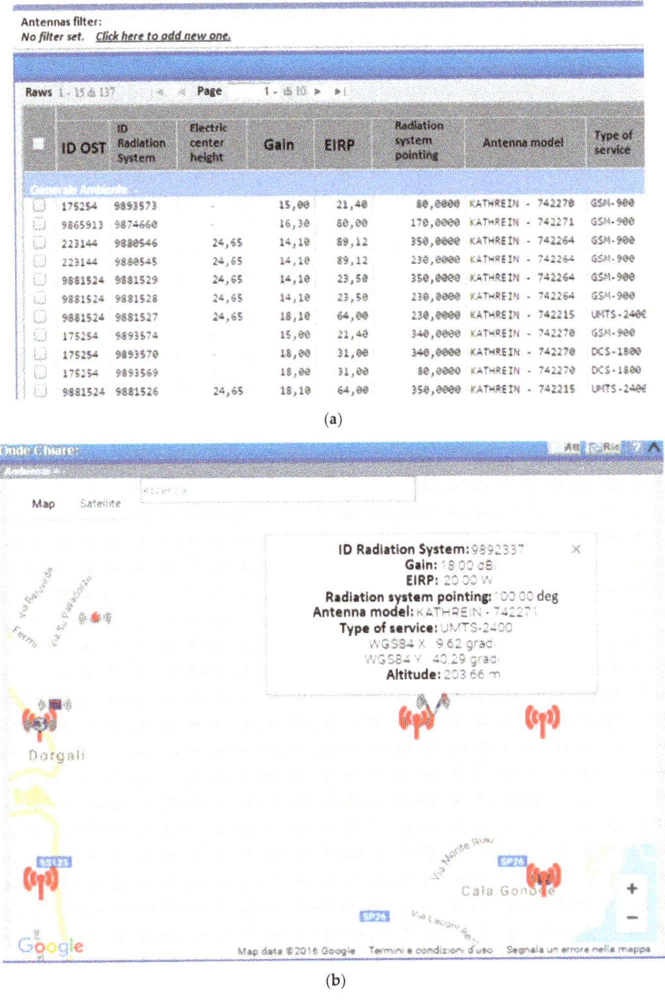

(a)

(b)

Figure 4. Examples of the web app of the Onde Chiare system. (**a**) The register of antennas contains information related to the position, the identification number, the height, the gain, the nominal power, the pointing direction, the antenna type, and the working frequency band. (**b**) The web app provides the actual location of the RBS in the region of interest, with the full set of information.

3.2. Onde Chiare App

The Onde Chiare app is a prototype designed for Android devices that was tested internally at the University of Cagliari [28]. Figure 5a shows the app main screen; it offers the possibility for each mobile device owner to become part of a network of distributed information made up of citizens interested in environmental issues. Indeed, the active participation of citizens has a fundamental role. The Onde Chiare app is an application that enables the community to promote a proper form of active involvement of citizens and real-time information sharing of electromagnetic field levels in a given geographical area. This mobile application provides some services such as:

1. Information about the geographic location and the input power of the antenna on the map, as shown in Figure 5b;
2. Geolocation of the electromagnetic field measurements;
3. Sending of geo-localized reports (i.e., broken antennas), as shown in Figure 5d.

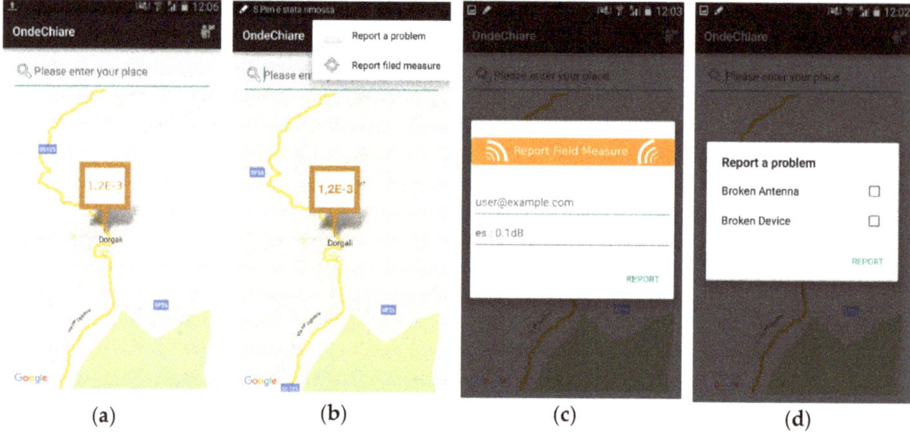

Figure 5. Examples of the app functionalities. (**a**) Main screen: the mobile user position is presented in the Google maps environment. (**b**) Report Menu. (**c**) User-suggested field measures for enhanced feedback. (**d**) Screen of the reported diagnostic tool.

The application allows a user to collect specific information directly from the system. Originally, we created our application in an Android development environment based on the Java Software Development Kit (SDK) 4 and Android 6.0. The application was developed using state-of-the-art frameworks, such as RoboGuice [29] for dependency injection [30] and Roboletric [31] for automated testing. Google Maps SDK was used to draw locations on a map in real time. The application places an informative marker on the user's current position on the map, as shown in Figure 5a. The user can also type a specific address, and the map shows all existing antenna masts. Each antenna mast is represented by an icon and by double-clicking on a specific antenna on the map, the basic technical information for the mast is displayed as in Figure 4b. The register includes information about the geographic location and the input power of the antenna and transmits to a centralized database via an unencrypted channel by stakeholders obliged to disclose information. This information is a set of organized technical data related to all transmission systems operating in the telecommunications sector, and the structured data are stored in a well-defined database. This system allows us to use our modified version of the COST 231 Walfisch–Ikegami model to estimate the field level in a given point within an urban area of interest. The OCS can be considered an effective tool for supporting regional actions in terms of planning, control, and supervision of the entire telecommunications system, while enabling us to assess the adherence to national obligations from the EM exposure point of view.

4. Model and System Validation: Measurement Campaign

To demonstrate that the modified version of the C231WI model can deal with the built-up scenarios of hilly, largely variable, and small, irregularly arranged towns, such as those often encountered in Italy, an experimental validation was carried out. The experimental validation is fundamental in order to use the proposed model to develop the software system.

The measurements were performed in the three UHF bands of interest. In particular, in the town of Dorgali (NU), the electric field levels were measured in the GSM band, at 944.2 MHz in the set of points shown in Figure 6a. For the frequency of 1847.8 MHz, the site of Cala Gonone (NU), presented in Figure 6b, was considered. Finally, the 4G coverage (2142.4 MHz) and exposure were assessed in the town of Lunamatrona (CA) in Sardinia (Italy), as shown in Figure 6c. Regarding the highest frequency band, a remark is in order. The C231WI model is known to be limited to an upper frequency bound of 2 GHz. This is a nominal constraint, but, in this work, relying on the method of [24], we corroborated the possibility of using the C231WI model at 2150 GHz with reasonable accuracy. At each site, the measurements were carried out using a YAGI antenna for both the 1.8 GHz and 2.15 GHz bands. On the other hand, a log-periodic antenna (LPDA) was used for the 900 MHz band. A tripod was used to position the antenna 1.5 m above ground. A Rohde-Schwarz FSH8 spectrum analyzer, operating from 9 kHz to 8 GHz and with an input impedance of 50 Ohm, was used to measure the electric field. Given the daily periodic pattern of the transmitted RBS power [32,33], which follows the traffic load and present variations of about 8–9 dBm between day and night [34], all the measurements were performed during the traffic peak hours [32]. At least 15 measurement points were selected for each location, and their position, as well the RBS location.

Figure 6. Topology of the three sites for the measurement campaigns. (**a**) Dorgali (NU, Italy): the working frequency of the RBS is 944.2 MHz, for a Kathrein 730376 antenna located 20 m above the road level. Fifteen measurement points were selected (A–Q). (**b**) Cala Gonone (NU, Italy): the working frequency of the RBS is 1878.4 MHz, for a Kathrein 742212 antenna located 10 m above the road level. Seventeen measurement points were selected (A–S). (**c**) Lunamatrona (CA, Italy): the working frequency of the RBS is 2142.4 MHz, for a Kathrein 742212 antenna located 30 m above the road level. Eighteen measurement points were selected (A–T).

5. Results

The extended version of the C231WI model [24] and the Onde Chiare system were experimentally validated, and their effectiveness in the estimation of electric field levels was assessed, as described in Section 4. The first urban scenario was the town of Dorgali (NU, Italy), as shown in Figure 6a. The measured electric field values are reported in Table 2. To derive the related estimated EM field levels, for each of the fifteen measurement points, using the new definition of the model parameters given in Section 2 (see Figure 2), the OCS is accessed to retrieve the information about the nearest RBS. Then, the 3D pattern of the antenna gain is derived using the reconstruction algorithm [25,26]; hence by applying Equation (1), the resulting theoretical electric field value is derived, as reported in Table 2. It can be noticed that at the frequency of 944.2 MHz (2G), for distances from the RBS which range from 48.5 m to 166.8 m, the maximum error is about 4.5 dB V/m, which is coherent with the original version of the model [4] and previous literature results [25]. When the frequency increases to 1847.4 MHz (3G), in the case of Cala Gonone (NU, Italy), i.e., Figure 6b, the Onde Chiare application, following the aforementioned steps, returns estimated values of the electric field levels very close to the measured ones, as reported in Table 3. Indeed, for a range of distances from the RBS greater than the previous case (i.e., from 62 m to 314.8 m) the average error is about 0.61 dB V/m, with a maximum error 10% lower than that found for 944.2 MHz. For the highest frequency band, i.e., 2142.4 MHz, at the site of Lunamatrona (CA, Italy) presented in Figure 6c, the modified C231WI model and the OCS can provide an estimated electric field level very similar (from −4.8 dB V/m to 4.5 dB V/m, for RBS distances from 412.7 m to 779 m) to the measured one, as shown in Table 4.

Table 2. Results of the Dorgali measurement campaign (944.2 MHz, Kathrein 730376).

Point	Distance from the RBS (m)	Measured (dB V/m)	Estimated (dB V/m)	Error (dB V/m)
A	48.5	−32.1	−28.1	4
B	60	−42	−44.1	−2.1
C	71.4	−29.6	−33.1	−3.5
D	67.6	−40.9	−44.2	−3.3
E	166.8	−24.4	−21.1	3.3
F	166	−41.3	−38.5	2.8
G	188	−40.9	−37.2	3.7
H	135	−40.7	−38.2	2.5
I	131.5	−26.9	−30.1	−3.9
L	88.4	−27	−28.7	−1.7
M	145	−40.5	−43.8	−3.3
N	115	−36.4	−36.6	−0.2
O	219	−27.2	−24.3	2.9
P	109	−38.6	−34.2	4.4
Q	119	−25.8	−21.6	4.2

Table 3. Results of the Cala Gonone measurement campaign (1847.8 MHz, Kathrein 742212).

Point	Distance from the RBS (m)	Measured (dB V/m)	Estimated (dB V/m)	Error (dB V/m)
A	79.8	−47.2	−44.8	2.4
B	62	−59.6	−59.1	0.5
C	101.5	−48.1	−44.6	3.5
D	140	−49.7	−46.9	2.8
E	183	−55	−53.9	1.1
F	173	−50.8	−52.6	−1.8
G	203.5	−50.5	−48.6	1.9
H	228	−57.2	−58.5	−1.3
I	273.5	−56.1	−59.4	−3.3
L	281.5	−63.4	−59.7	3.7
M	300	−52.1	−53.6	−1.5
N	148.5	−48.8	−45.3	3.5
O	150.6	−55.9	−60.2	−4.3
P	176	−49.3	−51.2	−1.9
Q	248.8	−50.4	−51.7	−1.3
R	282.5	−54.4	−51	3.4
S	314.8	−54.3	−51.3	3

In order to highlight that the new definitions of the model parameters which describe the set of buildings involved in the propagation between the RBS and the mobile are pivotal for the case of small, hilly towns with a significant variability in the topology and environment, it is worth noting that an accuracy equivalent to the one claimed by the original C231WI in the case of large and almost homogenous cities is obtained [4]. In other words, with the results from Table 2 to 4, it is demonstrated that the modified version of the C231WI model and the OCS perform similarly to the original model, but for a worst and complex case. We have summarized these findings in Figure 7 in order to present the model and system performances against the distance from the RBS and the working frequency.

The error of the estimated field level oscillates randomly with the distance from the RBS. However, from Figure 7 it can be noticed that for the three sites, the distribution of the error is similar for the three built-up areas of interest. This implies that the prediction error is similar across the UHF frequency band considered in this work. By remembering that the original C231WI model was developed to work at a maximum frequency of 2 GHz [4,25], the findings from Figure 7 corroborate the possibility of extending the model to higher frequency as long as its parameters are modified, as explained in Section 2. As a conclusion, the validation campaign demonstrated that the modified version of the C231WI model and the Onde Chiare system are effective, accurate, and reliable tools for the prediction of EM field levels in irregular, hilly built-up areas, thus being a technological solution for companies from the telecommunication system or for local regulatory agencies.

Table 4. Results of the Lunamatrona measurement campaign (2142.4 MHz, Kathrein 742212).

Point	Distance from the RBS (m)	Measured (dB V/m)	Estimated (dB V/m)	Error (dB V/m)
A	438	−53.3	−49.4	3.9
B	437.2	−59.3	−57.3	2
C	412.7	−53.5	−54.3	−0.8
D	394.5	−32.9	−28.3	4.6
E	483.4	−47.3	−44.7	2.6
F	519.5	−59.2	−54.7	4.5
G	478	−40.6	−45.4	−4.8
H	489	−40.5	−42.8	−2.3
I	526	−46.6	−48.1	−1.5
L	519.8	−47.6	−47.7	−0.1
M	499	−52.4	−48,.1	4.3
N	516	−55.7	−55.7	0
O	540	−44.3	−40.6	3.7
P	547	−43.3	−45	−1.7
Q	587	−44.3	−47.3	−3
R	698.3	−49.6	−51.5	−1.9
S	741	−58.7	−55.2	3.5
T	779	−60.1	−56.9	3.2

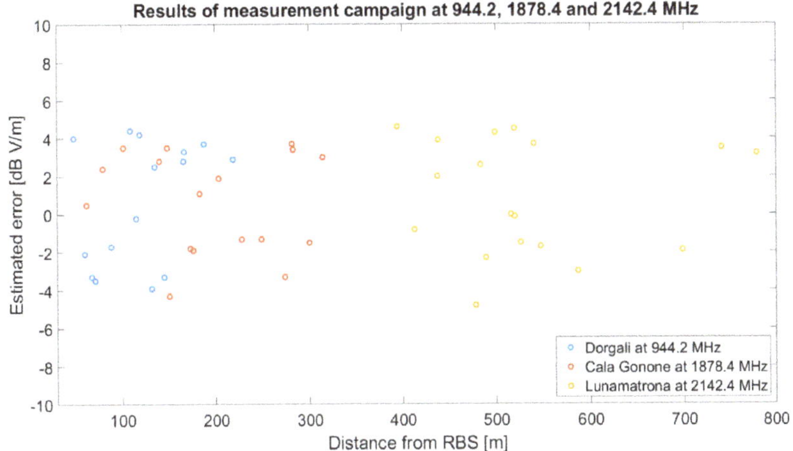

Figure 7. Comparison of the estimated error with our modified version of the C231WI model for the three built-up scenarios of Dorgali (NU, Italy), Cala Gonone (NU, Italy), Lunamatrona (CA, Italy) at the UHF frequencies of 944.2 MHz, 1878.4 MHz and 2142.4 MHz.

6. Conclusion and Future Work

In line with our research work, we addressed the problem of providing an effective, reliable, and quantitative model for the estimation of electromagnetic field levels in built-up areas. In particular, the COST 231 Walfisch–Ikegami model was re-phrased and modified in order to allow the description of small, hilly towns with buildings of largely variable height, arranged in irregular grids. The novel methodology was validated by performing a measurement campaign in three different sites for the three different GSM, 3G, and 4G frequencies, namely, 944.2 MHz, 1878.4 MHz, 2142.4 MHz. The values of the estimated electric field differed by a maximum of about 5 dB relative to measured ones, for a large set of distances from the RBS and for all frequencies. These findings corroborate the possibility of using the modified C231WI model for urban environments which are significantly different from those for which it was initially developed [25]. Furthermore, our results support the feasibility of extending the use of this model up to 2.5 GHz with reasonable accuracy. Given this experimental evidence, and since the development of 5G technology already resulted in the deployment of RBS stations, it is questionable if our modified C231WI model could be extended to the 3.6–3.8 GHz band, for both monitoring and coverage issues [35–37]. Indeed, in the literature, there is a lack of models and experimental strategies for EM propagation in urban environments [38], such as those investigated in this work, even though several studies which cover the synthesis of antennas, protocols, and systems for 5G exist [39–41]. Future work may deal with the definition and correction of the model parameters for EM signals of higher frequency, with the goal of ensuring the same accuracy and reliability of the estimated electric field levels.

Moreover, our research work aimed to answer the information needs of citizens about EM fields while satisfying the requirement of lowering the monitoring efforts of regulatory and local agencies. Accordingly, we developed a pre-industrial software prototype called the Onde Chiare System oriented to provide real-time knowledge of electric field levels in a given area. It is a valuable tool to improve the communication efforts by local authorities and facilitate the policy makers to make careful planning decisions and inform the citizens on environmental issues.

This technological solution may be further refined to develop a commercial product that could be of interest for telecommunication industries to speed-up the coverage assessment or setup a user-driven diagnostic system based on the app feedbacks. Furthermore, the Onde Chiare system may be used as a platform for regulatory agencies and entities for the environmental monitoring of general public exposure to electromagnetic radiation, while serving as an informative tool for citizens concerned about EM pollution-connected risks.

Author Contributions: Conceptualization, A.F., G.M.; methodology, L.S., M.B.L., A.F., G.M.; software, F.L., K.M., M.O.; validation, L.S., A.F., K.M., M.O., F.L; formal analysis, L.S., A.F., G.M.; investigation, L.S., M.B.L., A.F., G.M.; resources, L.S., F.L., M.B.L., A.F.; data curation, L.S., M.B.L.; writing—original draft preparation, L.S., M.B.L., K.M.; writing—review and editing, L.S., F.L., M.B.L., A.F., K.M., G.M.; visualization, L.S., M.B.L.; supervision, A.F., G.M., K.M. All authors have read and agreed to the published version of the manuscript.

Funding: This research was partially funded by the Ministero dello Sviluppo Economico, project "MONItoraggio distribuito sicuro, affidabile ed intelligente su tecnologie 5G: applicazione alla mobilità ed al servizio idrico", (MONIFIVE, CODICE CUP: F94I20000100001. This work was also supported by the Regione Autonoma Sardegna, under the grant for project Onde Chiare, P.O.R FESR SARDEGNA 2007–2013 ASSE VI COMPETITIVITA LINEA DI ATTIVITA 6.2.2 "SOSTEGNO ALLA CREAZIONE E SVILUPPO DI NUOVE IMPRESE INNOVATIVE." Asse I Ricerca, Sviluppo Tecnologico ed Innovazione; Determinazione RAS n. 33073 rep. N. 684 del 22/12/2014, CUP: E68C14000200007.

Acknowledgments: The authors would like to sincerely thank the staff of Oligamma IT S.r.l for the support for the software development.

Conflicts of Interest: The authors declare no conflict of interest.

References

1. Prasad, M.V.S.N.; Gupta, S.; Gupta, M.M. Comparison of 1.8 GHz Cellular Outdoor Measurement with AWAS Electromagnetic Code and Conventional Models Over Urban and Suburban Regions of Northern India. *IEEE Antennas Propag. Mag.* **2001**, *53*, 76–85. [CrossRef]
2. International Commission on Non-Ionizing Radiation Protection. Guidelines for Limiting Exposure to Electromagnetic Fields (100 kHz to 300 GHz). *Health Phys.* **2020**, *118*, 483–524. [CrossRef] [PubMed]

3. World Health Organization (WHO). *Establishing a Dialogue on Risks from Electromagnetic Fields*; WHO: Geneva, Switzerland, 2002.

4. COST Action 231. *Digital Mobile Radio towards Future Generation*; European Commission: Brussels, Belgium, 1999.

5. Bertoni, H.L. *Radio Propagation for Modern Wireless Systems*; Prentice & Hall: Upper Saddle River, NJ, USA, 1999.

6. Anglesio, L.; Benedetto, A.; Bonino, A.; Colla, D.; Martire, F.; Saudino Fusette, S.; d'Amore, G. Population Exposure to Electromagnetic Fields Generated by Radio Base Stations: Evaluation of the Urban Background by Using Provisional Model and Instrumental Measurements. *Radiat. Prot. Dosim.* **2001**, *97*, 355–358. [CrossRef] [PubMed]

7. Giliberti, C.; Boella, F.; Bedini, A.; Palomba, R.; Giuliani, L. Electromagnetic Mapping of Urban Areas: The Example of Monselice (Italy), Monselice (Italia). *PIERS Online* **2009**, *5*, 56–60. [CrossRef]

8. Miclaus, S.; Bechet, P. Estimated and measured values of the radiofrequency radiation power density around cellular base stations. *Rom. J. Phys.* **2007**, *52*, 429.

9. Djuric, N.; Kljajic, D.; Kasas-Lazetic, K.; Bajovic, V. The Measurement Procedure in the SEMONT Monitoring System. *Environ. Monit. Assess.* **2014**, *186*, 1865–1874. [CrossRef]

10. Djuric, N. The SEMONT continuous monitoring of daily EMF exposure in an open area environment. *Environ. Monit. Assess.* **2015**, *187*, 191. [CrossRef]

11. Koprivica, M. Statistical analysis of electromagnetic radiation measurements in the vicinity of GSM/UMTS base station installed on buildings in Serbia. *Radiat. Prot. Dosim.* **2016**, *168*, 489–502. [CrossRef]

12. Kljajic, D.; Djuric, N. Comparative analysis of EMF monitoring campaigns in the campus area of the University of Novi Sad. *Environ. Sci. Pollut. Res.* **2020**, *27*, 1–16. [CrossRef]

13. Çerezci, O.; Kanberoğlu, B.; Yener, Ş.Ç. Analysis on trending electromagnetic exposure levels at homes and proximity next to base stations along three years in a city. *J. Environ. Eng. Landsc. Manag.* **2015**, *23*, 71–81. [CrossRef]

14. Pascuzzi, S.; Santoro, F. Exposure of Farm Workers to Electromagnetic Radiation from Cellular Network Radio Base Stations Situated on Rural Agricultural Land. *Int. J. Occup. Saf. Ergon.* **2015**, *21*, 351–358. [CrossRef]

15. Ojuh, O.D.; Isabona, J. Radio Frequency EMF Exposure Due to GSM Mobile Phones Base Stations: Measurements and Analysis in Nigerian Environment. *Niger. J. Technol.* **2015**, *34*, 809–814. [CrossRef]

16. Saravanamuttu, S.; Singh, V.; Khumukcham, R.; Dorairaj, S. A Case Study of Cellular Base Stations in an Indian Metro (Chennai). *Environ. We Int. J. Sci. Technol.* **2015**, *10*, 37–49.

17. Zheng, Q.; Zhigang, W. Study on electromagnetic radiation tests of base stations and its influence scope. In Proceedings of the 2015 7th Asia-Pacific Conference on Environmental Electromagnetics (CEEM), Hangzhou, China, 4–7 November 2015.

18. Franceschetti, G.; Guida, R.; Iodice, A.; Riccio, D.; Ruello, G. Verifica di un Software per la Previsione Della Radiocopertura. In Proceedings of the XVI RiNEm, Genova, Italy, 18–21 September 2006.

19. *Telecomunicazioni ALDENA Software*. Programma software ALDENA per ambiente Windows®per la previsione dei campi elettromagnetici generati da antenne trasmittenti. 2007. Available online: http://www.aldena.it/ (accessed on 5 May 2020).

20. Berg, J-E. A recursive method for street microcell path loss calculations. In Proceedings of the 6th International Symposium on Personal, Indoor and Mobile Radio Communications, Toronto, ON, Canada, 27–29 September 1995.

21. Hata, M. Empirical formula for propagation loss in land mobile radio services. *IEEE Trans. Veh. Technol.* **1980**, *29*, 317–325. [CrossRef]

22. Okumura, Y. Field strength and its variability in VHF and UHF land-mobile radio service. *Rev. Electr. Commun. Lab.* **1968**, *16*, 825–873.

23. Walfisch, J.; Bertoni, H.L. A theoretical model of UHF propagation in urban environments. *IEEE Trans. Antennas Propag.* **1988**, *36*, 1788–1796. [CrossRef]

24. Ikegami, F.; Yoshida, S.; Takeuchi, T.; Umehira, M. Propagation Factors Controlling Mean Field Strength on Urban Streets. *IEEE Trans. Antennas Propag.* **1984**, *32*, 822–829. [CrossRef]

25. Fanti, A.; Schirru, L.; Casu, S.; Lodi, M.B.; Riccio, G.; Mazzarella, G. Improvement and Testing of Models for Field Level Evaluation in Urban Environment. *IEEE Trans. Antennas Propag.* **2020**. [CrossRef]

26. D'Agostino, F.; Iacone, M.; Riccio, G. A new recursive approach for the full pattern reconstruction. In Proceedings of the International Symposium on Antenna Technology and Applied Electromagnetics, ANTEM 2004, Ottawa, ON, Canada, 20–23 July 2004.

27. ARPAS. Available online: http://www.sardegnaambiente.it/arpas/ (accessed on 1 April 2020).

28. Mannaro, K.; Ortu, M. Onde Chiare: A Mobile Application to Mitigate the Risk Perception from Electromagnetic Fields. In *Mobile Web and Intelligent Information Systems*; Springer: Berlin/Heidelberg, Germany, 2016; Volume 9847.

29. Roboguice. Available online: https://github.com/roboguice/roboguice (accessed on 1 April 2020).

30. García, B.; López-Fernández, L.; Gortázar, F.; Gallego, M. Practical Evaluation of VMAF Perceptual Video Quality for WebRTC Applications. *Electronics* **2019**, *8*, 854. [CrossRef]

31. Roboeletric. Available online: http://robolectric.org/ (accessed on 1 April 2020).

32. Koprivica, M.; Petric, M.; Popovic, M.; Milinkovic, J.; Neškovic, A. Empirical analysis of electric field strength long-term variability for GSM/DCS/UMTS downlink band. *Telfor J.* **2016**, *8*, 87–92. [CrossRef]

33. Mahfouz, Z.; Gati, A.; Lautru, D.; Wong, M.F.; Wiart, J.; Hanna, V.F. Influence of traffic variations on exposure to wireless signals in realistic environments. *Bioelectromagnetics* **2012**, *33*, 288–297. [CrossRef] [PubMed]

34. Garcia Sanchez, M.; Cuiñas, I.; Alejos, A.V. Electromagnetic field level temporal variation in urban areas. *Electron. Lett.* **2005**, *41*, 233–234. [CrossRef]

35. Simkó, M.; Mattsson, M.O. 5G Wireless Communication and Health Effects—A Pragmatic Review Based on Available Studies Regarding 6 to 100 GHz. *Int. J. Environ. Res. Public Health* **2019**, *16*, 3406. [CrossRef] [PubMed]

36. Xu, B. Power density measurements at 15 GHz for RF EMF compliance assessments of 5G user equipment. *IEEE Trans. Antennas Propag.* **2017**, *65*, 6584–6595. [CrossRef]

37. Gkonis, P.K.; Trakadas, P.T.; Kaklamani, D.I. A Comprehensive Study on Simulation Techniques for 5G Networks: State of the Art Results, Analysis, and Future Challenges. *Electronics* **2020**, *9*, 468. [CrossRef]

38. Franci, D.; Coltellacci, S.; Grillo, E.; Pavoncello, S.; Aureli, T.; Cintoli, R.; Migliore, M.D. An Experimental Investigation on the Impact of Duplexing and Beamforming Techniques in Field Measurements of 5G Signals. *Electronics* **2020**, *9*, 223. [CrossRef]

39. Matalatala, M. Multi-objective Optimization of Massive MIMO 5G Wireless Networks towards Power Consumption, Uplink and Downlink exposure. *Appl. Sci.* **2019**, *9*, 4974. [CrossRef]

40. Qamar, F.; Hindia, M.H.D.; Dimyati, K.; Noordin, K.A.; Majed, M.B.; Abd Rahman, T.; Amiri, I.S. Investigation of Future 5G-IoT Millimeter-Wave Network Performance at 38 GHz for Urban Microcell Outdoor Environment. *Electronics* **2019**, *8*, 495. [CrossRef]

41. Ganame, H.; Yingzhuang, L.; Ghazzai, H.; Kamissoko, D. 5G Base Station Deployment Perspectives in Millimeter Wave Frequencies Using Meta-Heuristic Algorithms. *Electronics* **2019**, *8*, 1318. [CrossRef]

 *electronics*

Article

Analysis of a Nonlinear Technique for Microwave Imaging of Targets Inside Conducting Cylinders

Alessandro Fedeli *, Matteo Pastorino, Andrea Randazzo and Gian Luigi Gragnani

Department of Electrical, Electronic, Telecommunications Engineering, and Naval Architecture, University of Genoa, 16145 Genoa, Italy; matteo.pastorino@unige.it (M.P.); andrea.randazzo@unige.it (A.R.); gianluigi.gragnani@unige.it (G.L.G.)
* Correspondence: alessandro.fedeli@unige.it

Abstract: Microwave imaging of targets enclosed in circular metallic cylinders represents an interesting scenario, whose applications range from biomedical diagnostics to nondestructive testing. In this paper, the theoretical bases of microwave tomographic imaging inside circular metallic pipes are reviewed and discussed. A nonlinear quantitative inversion technique in non-Hilbertian Lebesgue spaces is then applied to this kind of problem for the first time. The accuracy of the obtained dielectric reconstructions is assessed by numerical simulations in canonical cases, aimed at verifying the dependence of the result on the size of the conducting enclosure and comparing results with the conventional free space case. Numerical results show benefits in lossy environments, although the presence and the type of resonances should be carefully taken into account.

Keywords: inverse scattering; tomography; conducting cylinders

Citation: Fedeli, A.; Pastorino, M; Randazzo, A.; Gragnani, G.L. Analysis of a Nonlinear Technique for Microwave Imaging of Targets Inside Conducting Cylinders. *Electronics* 2021, 10, 594. https://doi.org/10.3390/electronics10050594

Academic Editor: Reza K. Amineh

Received: 2 February 2021
Accepted: 27 February 2021
Published: 4 March 2021

Publisher's Note: MDPI stays neutral with regard to jurisdictional claims in published maps and institutional affiliations.

1. Introduction

Since the end of the past century [1], inverse electromagnetic scattering problems aimed at microwave imaging have experienced a widespread interest among researchers. Indeed, while still at a prototyping level in many applications, microwave imaging is evolving to a mature technique in many fields of civil engineering, industrial testing, medicine, and so on [2,3]. Although a comprehensive review about the topic is outside the scope of the present paper, the interested reader is referred to [4–7], where many of the theoretical aspects of microwave imaging as well as a variety of practical applications are considered and discussed, and a wide literature is suggested.

Among the various possible configurations of the microwave imaging problem, a scenario that is acquiring an increasing interest is represented by targets enclosed inside conducting enclosures [8,9]. Indeed, this kind of configuration may be useful in systems for biomedical diagnostics [10–13], as well as to inspect the content of pipes [14] or storage containers [15] in industrial processes and agriculture.

On the one hand, the use of conductive enclosures shields the investigation and measurement domains from external interferences and provides well-defined boundary conditions. Moreover, the container bounded by the outer conductor may be filled with proper (liquid or gel) matching media, which are particularly useful in medical applications. On the other hand, the electromagnetic (EM) field is usually strongly perturbed by the presence of the outer conductor, which should be taken into account inside the inverse scattering algorithm.

In this paper, we consider the cylindrical configuration schematized in Figure 1. In particular, we are interested in conductive circular enclosures, which can be part of the tomographic imaging system themselves, as for example it is in some devices for breast or head imaging, or can stand for the boundary of the domain to be imaged, as it could be in the case of monitoring inside pipelines.

107

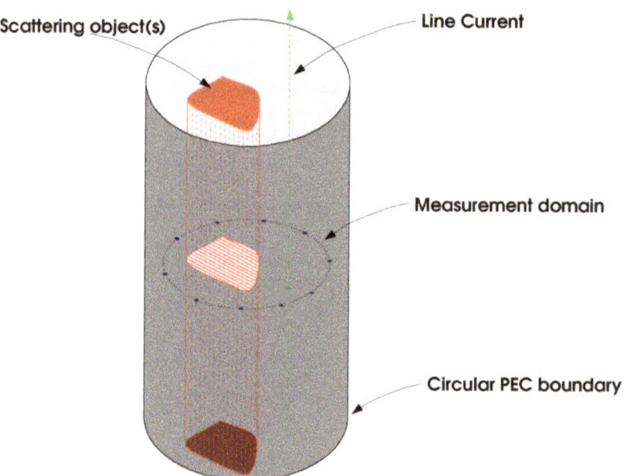

Figure 1. Schematic geometry of the considered problem.

As to the solution of microwave inverse scattering problems, innovative techniques are continuously proposed by the research community [16–22]. It is also worth noting that some specific approaches have been investigated for the imaging inside metallic enclosures, such as eigenfunction-based inversion algorithms [23,24]. Among the various possible inversion schemes, Newton-based methods look promising in many contexts, for their effectiveness in dealing with the intrinsic nonlinearity of the problem at hand [25–27]. Furthermore, it has been shown that facing the problem in non-Hilbertian Lebesgue spaces [28–30] may allow a more accurate reconstruction of the dielectric properties of the imaged domain in various configurations [31,32].

In this paper, we investigate the possibility of exploiting a Newton-based inversion method in non-Hilbertian Lebesgue spaces [33] for the microwave imaging of targets located inside conducting cylinders, illuminated by antennas modeled as line-current sources in the axial direction. In particular, only perfectly electric conducting (PEC) enclosures are taken into account. The adoption of the configuration of Figure 1 reflects in a change of the Green's function and consequently of the kernel of the integral operators used to describe the electromagnetic problem, with non-negligible effects on results. The proposed inversion technique is assessed with a set of numerical simulations in canonical cases, aimed at verifying the dependence of the imaging performance on the size of the enclosure and the loss of the infill dielectric medium, also comparing results with the free space case. Furthermore, for a fixed size of the imaging chamber, a comparison with conventional L^2 regularization is presented.

The paper is organized as follows. The electromagnetic scattering problem is formulated in Section 2, highlighting the specific issues associated with the presence of a conducting enclosure. Section 3 briefly outlines the structure of the adopted inverse scattering method. Numerical results are described and discussed in Section 4. At the end of the paper, Section 5 draws some concluding considerations.

2. Problem Formulation

In this paper, two-dimensional (2D) configurations are assumed, where the geometry is invariant along the z axis. Under z-polarized transverse magnetic (TM) illumination conditions, all the involved quantities only depend on the position in the transverse plane, which is indicated by cylindrical coordinates ρ, ϑ as defined in Figure 2. Time-harmonic electromagnetic fields are assumed with a time dependence $e^{j\omega t}$. Nonmagnetic materials are considered, characterized by a permeability $\mu = \mu_0 = 4\pi \times 10^{-7}$ H/m. The scattering object(s) under test (OUT) are placed inside a circular conducting cylinder of internal radius a, which is centered at the axes origin. The cylindrical enclosure is filled by a background medium with a possibly complex dielectric permittivity $\varepsilon_b = \varepsilon_0 \tilde{\varepsilon}_{r,b} = \varepsilon_0 \varepsilon_{r,b} - j\sigma_b/\omega$, where

$\varepsilon_0 \simeq 8.85 \times 10^{-12}$ F/m is the vacuum dielectric permittivity, $\varepsilon_{r,b}$ is the relative permittivity, and σ_b is an (equivalent) background conductivity. Illumination field is provided by an ideal infinite line of current of complex amplitude I_z placed at (ρ_i, ϑ_i), i.e.:

$$\bar{J}(\rho, \vartheta) = \frac{I_z}{\rho}\delta(\rho - \rho_i)\delta(\vartheta - \vartheta_i)\hat{z} \tag{1}$$

where $\frac{1}{\rho}\delta(\rho - \rho_i)\delta(\vartheta - \vartheta_i)$ is the Dirac δ-function for cylindrical coordinates and z invariance [34].

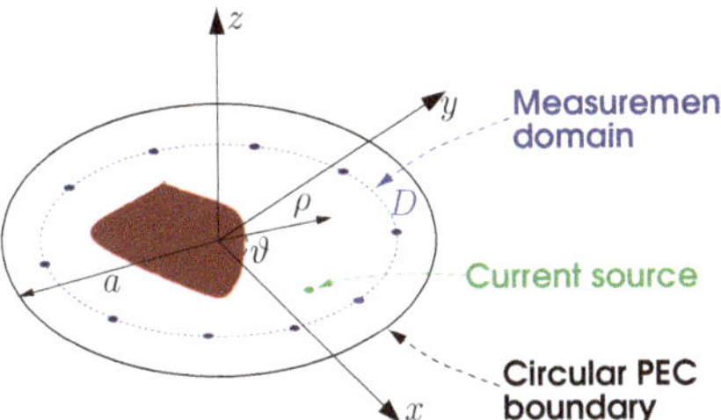

Figure 2. Assonometric projection of the transverse plane, and the related coordinate systems.

Let us denote as $e_i(\rho, \theta)$ the z component of the incident electric field inside the cylinder, which is a solution to the scalar Helmholtz equation:

$$\nabla^2 e_i(\rho, \vartheta) + k_b^2 e_i(\rho, \vartheta) = j\omega\mu_0 \frac{I_z}{\rho}\delta(\rho - \rho_i)\delta(\vartheta - \vartheta_i) \tag{2}$$

where $k_b = \omega\sqrt{\mu_0 \varepsilon_b}$ is the wavenumber, subject to the boundary condition $e_i(a, \theta) = 0$.

If some scattering object is present inside the cylinder, the (z-directed) total electric field $e_t(\rho, \theta)$ can always be written as the sum:

$$e_t(\rho, \theta) = e_i(\rho, \theta) + e_s(\rho, \theta) \tag{3}$$

where $e_s(\rho, \theta)$ is the (z-directed) scattered field.

According to the volume equivalence theorem [35], the scattering field can be expressed in an integral form

$$e_s(\rho, \theta) = -k_b^2 \int_D \tau(\rho', \vartheta')e_t(\rho', \vartheta')g(\rho, \vartheta; \rho', \vartheta')d\rho'd\vartheta' \tag{4}$$

where D is the considered investigation domain (which as reported in Figure 2, is a circular region located in the transverse plane) and

$$\tau(\rho, \vartheta) = \frac{\tilde{\varepsilon}_r(\rho, \vartheta)}{\tilde{\varepsilon}_{r,b}} - 1 \tag{5}$$

is the so-called object function, accounting for the dielectric properties of the scatterers by means of the space-dependent complex relative dielectric permittivity $\tilde{\varepsilon}_r(\rho, \vartheta)$, and finally $g(\rho, \vartheta; \rho', \vartheta')$ is the Green function for the problem.

Green's Function of the Considered Problem

The Green's function inside a circular PEC cylinder of radius a (Figure 1) is the solution of the following Helmholtz equation:

$$\nabla^2 g(\rho, \vartheta; \rho', \vartheta') + k_b^2 g(\rho, \vartheta; \rho', \vartheta') = \frac{1}{\rho}\delta(\rho - \rho')\delta(\vartheta - \vartheta') \tag{6}$$

subject to the boundary condition $g(a, \vartheta; \rho', \vartheta') = 0$.

The solution to this equation is discussed in many articles and textbooks (see, for example [35–38]), and can be expressed in some different forms. In this work the following expression will be used [35]:

$$g(\rho, \vartheta; \rho', \vartheta') = -\frac{1}{4}\begin{cases} \sum\limits_{n=-\infty}^{+\infty} \left[J_n(k_b\rho')Y_n(k_ba) - J_n(k_ba)Y_n(k_b\rho') \right] \times \\ \qquad \times \dfrac{J_n(k_b\rho)}{J_n(k_ba)} e^{jn(\vartheta-\vartheta')} \qquad\qquad\qquad \rho < \rho' \\[2mm] \sum\limits_{n=-\infty}^{+\infty} \left[J_n(k_b\rho)Y_n(k_ba) - J_n(k_ba)Y_n(k_b\rho) \right] \times \\ \qquad \times \dfrac{J_n(k_b\rho')}{J_n(k_ba)} e^{jn(\vartheta-\vartheta')} \qquad\qquad\qquad \rho > \rho' \end{cases} \tag{7}$$

where J_n and Y_n are the Bessel functions of the first and of the second kind, respectively, of order n. Without going in too much theoretical details (which were the subject of other works, such as [8]) here we want just to stress two aspects of the problem, making it much more tricky than the analogous in free-space:

- while in free space the scattering field can be expanded into a simple sum of progressing waves, in the present problem the solution is made by a sum of complicated standing waves, and many resonant modes can arise inside the cavity;
- the incident field is also strongly affected by the cavity boundaries: while the line current produces a simple circular wave in free space, in the present problem the incident field has the same form of the Green's function (compare Equations (2) and (6)), hence it contains many contributions, made of standing waves depending on the cavity dimensions.

In (7), the presence of the term $J_n(k_ba)$ causes g to assume high values when it approaches zero (in particular, $g \to \infty$ when $J_n(k_ba) = 0$). In the lossless case, the positions of TM_{nl} resonances can be found based on the l-th root of the Bessel function of first kind with order n.

Investigation about the zeros of Bessel's functions is a problem that is not new in physics and the first comprehensive works about this topic date back the second half of the 19th century [39,40]. Many of these and other findings (e.g., [41]) about Bessel's functions are collected and extended in the monumental book by Watson [42]. Further studies were since then carried out and this topic is presently still an open research field for mathematics. A complete review of the literature about Bessel's functions is outside the scope of the present paper; however we refer the reader to some papers [43–53], containing relevant results for the problem faced in the present work. It is worth nothing that also some works, aimed at other goals, contain notable insight about the zeros of Bessel's functions [54–58].

In particular, in order to underline the behavior of resonances within the enclosure, some properties of the zeros of Bessel's functions are worth recalling. Please note that such properties can be valid for any real order v, but we focus on integer order n. Let $\zeta_{n,l}$ denote the l-th ($l \geq 1$) zero of the Bessel function of first kind with order n. Then:

a. $\lim_{n\to\infty}(\zeta_{n,l+1} - \zeta_{n,l}) = \pi$;

b. $\zeta_{n,1} < \zeta_{n+1,1} < \zeta_{n,2} < \zeta_{n+1,2} < \zeta_{n,3} < \ldots$;

c. $\zeta_{n,l} > n + l\pi + \dfrac{\pi}{2} - \dfrac{1}{2}$;

d. $\zeta_{n,l} < \left(l + \dfrac{n}{2} - \dfrac{0.965}{4}\right)\pi - \dfrac{n^2}{2}\left[\left(l + \dfrac{n}{2} - \dfrac{0.965}{4}\right)\pi\right]^{-1} \quad n > 0$;

e. $\zeta_{n,1} < 2(n+1)(n+5)(5n+11)/(7n+19) \quad n > -1$.

The first property states that, for each n, the zeros approaches a regular distribution, in accordance with the asymptotic behavior of J_n. The second statement is about the interlacing property of the zeros. This should explain how the zeros of Bessel's functions of any order tend to distribute. In particular, it should be stressed that between two consecutive zeros of J_n lies exactly one zero of J_{n+1}. Furthermore, property (c) provides a lower bound for the zeros, and it can be seen that "zeros are strictly increasing functions of both index and order" [58]. Therefore, properties (a), (b), and (c) show that the number of zeros in an interval $(x - \Delta, x + \Delta)$ strictly increases as x becomes larger.

The number of zeros in an interval could also be estimated by using upper bounds that exists for the zeros of Bessel functions. Properties (e) and (d) are about two simple upper bounds. More sharp (and complicated) bounds can also be found in the literature (e.g., [45,47,48,59,60]); however, for the problem faced in the present work, the number of zeros is small enough to allow computing to a high accuracy the position of each zero.

Previous analysis is strictly related to the problem dealt with in the present work, since, as the radius of the enclosure increases, the number of possible resonances increases, too, and their distribution follows the behavior of Bessel zeros. As an example, zeros corresponding to the resonances of a conducting enclosure filled with vacuum and with radius a, ranging between 0 and 3λ, have been evaluated, and their number have been computed in each of the intervals $(x - \Delta, x + \Delta)$, $\Delta = 0.25\lambda$, $x = [0.25 + m/2]\lambda$, $m = 0, \ldots, 5$. In Figure 3 the histogram of the results is shown. As can be seen, the predicted behavior is perfectly confirmed. For convenience, the first TM_{nl} resonances versus the radius of the PEC cylinder (up to 3λ) are reported in Table 1.

Table 1. First TM_{nl} resonances versus the radius of the conducting enclosure (for radiuses up to 3λ).

Order, n	Root, l	Radius, a/λ	Order, n	Root, l	Radius, a/λ
0	1	0.382565575636716	9	1	2.12443502732519
1	1	0.609557227746547	6	2	2.16181776676093
2	1	0.816987450864820	4	3	2.28641855021103
0	2	0.878147628239934	10	1	2.30279831942887
3	1	1.01497187625880	2	4	2.35377646939679
1	2	1.11605681508930	7	2	2.35780395133906
4	1	1.20717221350811	0	5	2.37524718150666
2	2	1.33903593944348	11	1	2.48007141762091
0	3	1.37665636071269	5	3	2.49762238062566
5	1	1.39538925994123	8	2	2.55132863821835
3	2	1.55280761241691	3	4	2.58086897429352
6	1	1.58066078745604	1	5	2.62018841505947
1	3	1.61842038016315	12	1	2.65639874693837
4	2	1.76020125084557	6	3	2.70500953319621
7	1	1.76364706145171	9	2	2.74277582388405
2	3	1.84851296702361	4	4	2.80239128808070
0	4	1.87582635499932	13	1	2.83189617009082
8	1	1.94479780216248	2	5	2.85709234145654
5	2	1.96285556025129	0	6	2.87478938633534
3	3	2.07049020258037	7	3	2.90923374085231
1	4	2.11956574517874	10	2	2.93244082347851

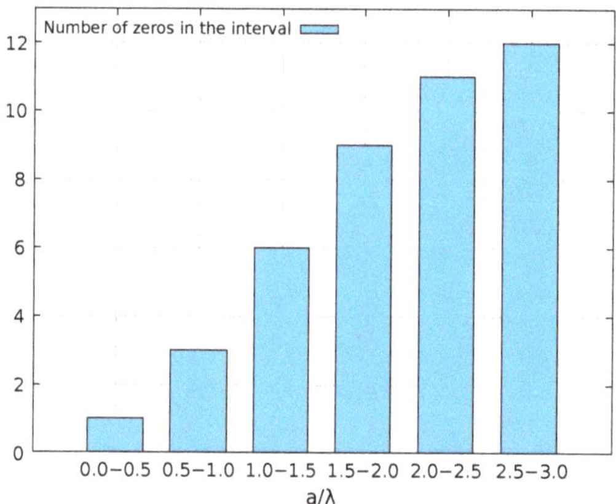

Figure 3. Number of resonances for an enclosure as the radius ranges from 0 to 3λ.

When the argument of a Bessel function becomes complex, as is in the case of conductive media, the behavior changes in accordance to the variation on the imaginary part $\Im\{z\}$ of the argument $z = x - jy$. In particular, while $J_n(x)$ is bounded, this is no longer the case for $J_n(x - jy)$, where both the real and the imaginary parts tend to increase as the argument increases. An in-depth discussion about Bessel's function of complex arguments is outside the scope of the present work and we refer interested readers to some specialized literature [61,62]. However, to investigate some practical case, let us consider:

$$z = x - jy = k_b\rho = [\Re\{k_b\} - j\Im\{k_b\}]\rho \tag{8}$$

where k_b is complex, while ρ is real. In particular, two effects can be observed: the behavior with respect to lossless case gets smoothed, since there are no longer zeros, and the amplitude of $J_n(k_b\rho)$ tends to oscillate and decrease up to a certain value of ρ. For larger values, oscillations are quickly damped and $|J_n(k_b\rho)|$ begins to increase and eventually approaches infinity. In Figure 4 an example is shown, for different orders of Bessel functions and for different values of σ_b. In particular, in the upper left corner, the function giving the larger number of zeros, namely J_0, is plotted, while in lower right corner we have shown the behavior of J_{13}, being 13 the maximum order for which it is possible having a zero in the considered range. The other two graphs, referring respectively to J_3 and J_7, are intended to give an idea of the trend, as the order of the Bessel function varies. From the example in Figure 4 it can be deduced that, while in general, a conductive background would be an obstacle to the solution of an inverse scattering problem, for a certain range of values of σ_b (we could consider $\sigma_b \lesssim 3$ mS/m in the example, with $\Im\{k_b\} \lesssim 0.15\Re\{k_b\}$) an advantage can come from the damping of the resonances, while the losses can still be managed.

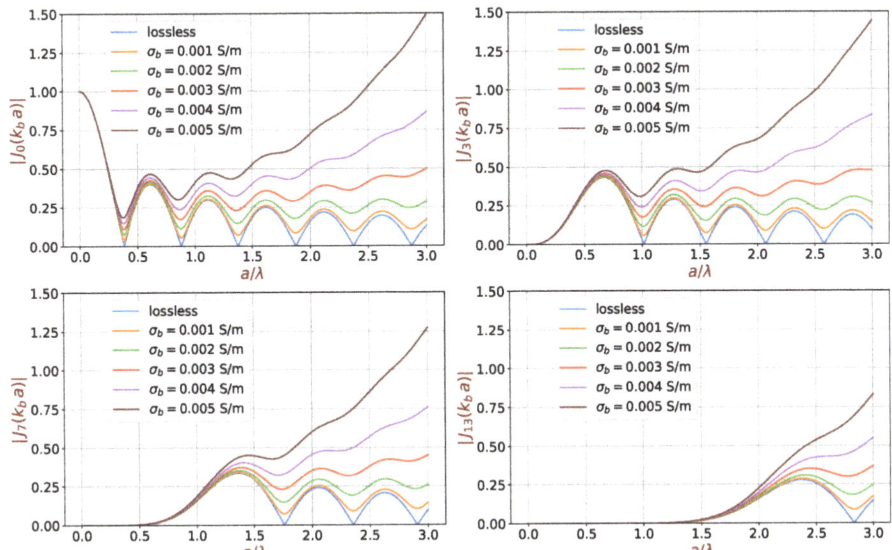

Figure 4. Behavior of the magnitude of different Bessel's functions $J_n(k_b a)$ for a background with $\varepsilon_b = \varepsilon_0 - j\sigma_b/\omega$ and different values of σ_b, versus the radius of PEC cylinder a. In the upper left corner J_0 is shown.

3. Nonlinear Inverse Scattering Method

In solving the inverse problem, the goal is to retrieve the object function $\tau(\rho, \vartheta)$ inside an investigation domain D starting from a set of measurements of the scattered field $e_s(\rho, \theta)$ collected in an observation domain O. Based on (4), it can be shown that $\tau(\rho, \vartheta)$ is connected to the measured scattered electric field $e_s(\rho, \theta)$ by means of a nonlinear relationship:

$$e_s(\rho, \theta) = \mathcal{A}_O \tau (I - \mathcal{A}_D \tau)^{-1} e_i(\rho, \theta) = \mathcal{N}(\tau)(\rho, \theta), \quad (\rho, \theta) \in O \tag{9}$$

where the linear operators $\mathcal{A}_{\{O,D\}}$ are given by

$$\mathcal{A}_{\{O,D\}} x(\rho, \theta) = -k_b^2 \int_D x(\rho', \vartheta') g(\rho, \vartheta; \rho', \vartheta') d\rho' d\vartheta', \quad (\rho, \theta) \in \{O, D\} \tag{10}$$

The inverse scattering method is characterized by an inexact-Newton scheme, where (9) is iteratively linearized around the reconstructed value of the object function $\tau_i(\rho, \vartheta)$ at the i-th step ($i = 1, ..., I$). Starting the iterations with an initial value $\tau_1 = 0$, the result of such a linearization is the equation

$$\mathcal{N}_i' \delta = \underbrace{e_s - \mathcal{N}(\tau_i)}_{b_i} \tag{11}$$

where $\delta \in A$, $b_i \in B$ and $\mathcal{N}_i' : A \to B$ is the Fréchet derivative of the nonlinear operator $\mathcal{N}$ calculated at τ_i. The solution of (11) in the unknown δ is again performed by an iterative loop, which is a truncated Landweber-like method that considers A, B as Lebesgue spaces L^p [33]. This regularized solution approach, at the k-th step ($k = 1, ..., K$), has been implemented as

$$\delta_{k+1} = J_{A^*}\left[J_A(\delta_k) - \gamma \mathcal{N}_i'^* J_B\left(\mathcal{N}_i' \delta_k - b_i \right) \right] \tag{12}$$

where $J_A(\cdot)$, $J_{A^*}(\cdot)$ and $J_B(\cdot)$ are defined as the duality maps of spaces A, A^* (i.e., the dual space of A) and B [31]; $\mathcal{N}_i'^*$ is the adjoint of $\mathcal{N}_i'$; γ is a real positive number. The loop starts with $\delta_1 = 0$. Once (11) is solved by means of iterations (12), a regularized version of the unknown of the linear problem $\tilde{\delta}$ is available. Then, the object function inside the investigation domain is updated as

$$\tau_{i+1}(\rho,\theta) = \tau_i(\rho,\theta) + \tilde{\delta}(\rho,\theta), \quad (\rho,\theta) \in D \tag{13}$$

and a new linearization is performed, until convergence is reached.

4. Results of Numerical Simulations

The behavior of the proposed inverse scattering technique when operating in conducting enclosures has been analyzed from a numerical point of view in canonical configurations, involving circular dielectric cylinders.

The simulations have been oriented at understanding the influence of the external conducting cylinder on the reconstruction performance. To this end, two distinct cases have been considered, regarding a small and a large enclosure. The difference between these two cases is related to the definition of the investigation domain D and the observation domain O. This distinction has also been made to follow, in the case of sufficiently large enclosures, the guidelines about the position of the observation domain with respect to the conducting boundary provided in [8]. A background characterized by $\varepsilon_b = \varepsilon_0 - j\sigma_b/\omega$ has been considered. The electric field data have been simulated at the angular frequency $\omega = 2\pi f$, with $f = 300$ MHz, by means of a custom numerical code based on the method of moments, and corrupted with an additive Gaussian noise with zero mean value and signal-to-noise ratio equal to 20 dB. In all cases, the inversion method has been run with the same parameters in order to compare results. In particular, a Lebesgue space exponent $p = 1.2$ is considered; the iterative loops have been stopped when a maximum number of inexact-Newton iterations $I = 10$ and a maximum number of Landweber steps $K = 100$ are reached, or when the relative variation of the data residual falls below a threshold value $r_{th} = 1\%$.

The results have been evaluated from a quantitative viewpoint by means of the normalized reconstruction error (NRE)

$$NRE = \frac{||\tau - \tau_{act}||}{||\tau_{act}||} \tag{14}$$

For each analyzed case, $N = 15$ simulations with different random noise have been executed, calculating the mean value and the standard deviation of the obtained values of the NRE.

4.1. Small Conducting Enclosure

In the case of the small conducting enclosure, the radius of the outer PEC cylinder has been varied in the range $a \in [0.25, 1.025]\lambda$ with 0.025λ steps. A set of $S = 30$ positions, equally spaced on a circumference with radius $\rho_s = a - 0.1\lambda$, have been defined to host both sources and measurement points in a multistatic and multiview configuration (i.e., one of the position at a time is occupied by a source antenna, whereas all the other ones are used to sample the scattered electric field).

The investigation domain is a circular region centered at the origin with diameter $d_D = 0.25\lambda$. This region contains a circular dielectric cylinder of diameter $d_c = 0.125\lambda$, centered at $(\rho_c, \theta_c) = (0.05\lambda, 0)$, characterized by a relative dielectric permittivity $\varepsilon_{r,c} = 2$ and electric conductivity $\sigma_c = 10$ mS/m. For the forward problem solution, D has been subdivided into $N_f = 1976$ square subdomains with side length equal to $l_{sf} = 0.005\lambda$. Conversely, a discretization with $N_i = 1264$ cells of side $l_{si} = 0.00625\lambda$ has been adopted inside the inversion procedure.

Figure 5 reports the NRE reconstruction error versus the size of the enclosure a, for three different values of background loss $\sigma_b = \{0, 1, 10\}$ mS/m (which correspond to loss tangents $\tan\delta = \{0, 0.06, 0.6\}$). Some examples of reconstructed dielectric permittivity are also shown in Figures 6 and 7, along with the magnitude of the Green's function inside the conducting cylinder for a source at $(\rho_s, 0)$. All results are compared with the free space case, where the same configuration but without the PEC enclosure have been simulated.

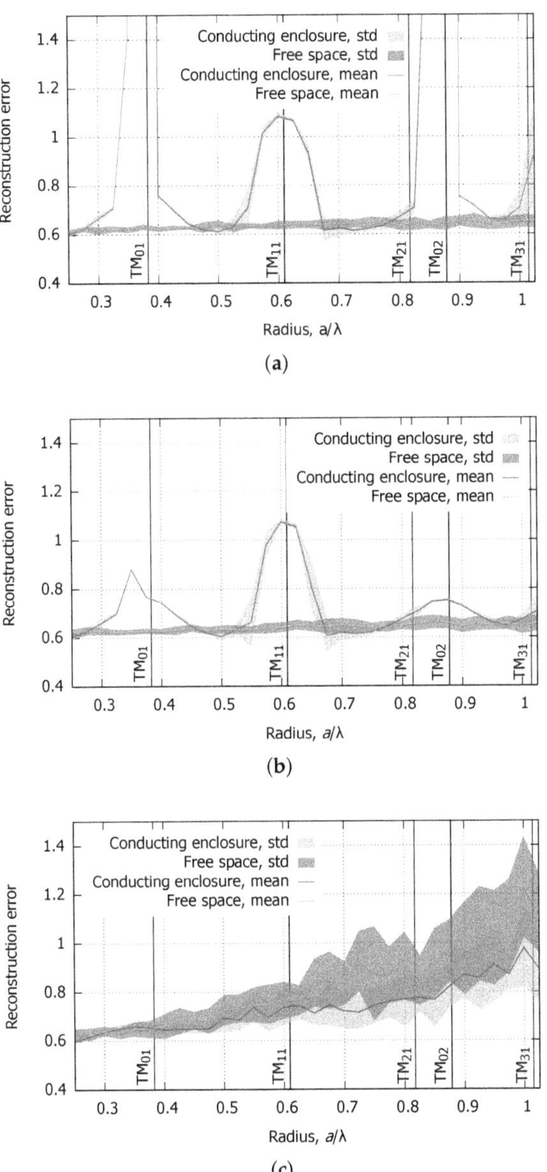

Figure 5. Small conducting enclosure. Reconstruction error versus the radius of the conducting enclosure, for three values of the background conductivity: (**a**) lossless, $\sigma_b = 0\,\mathrm{S/m}$; (**b**) $\sigma_b = 1\,\mathrm{mS/m}$; (**c**) $\sigma_b = 10\,\mathrm{mS/m}$.

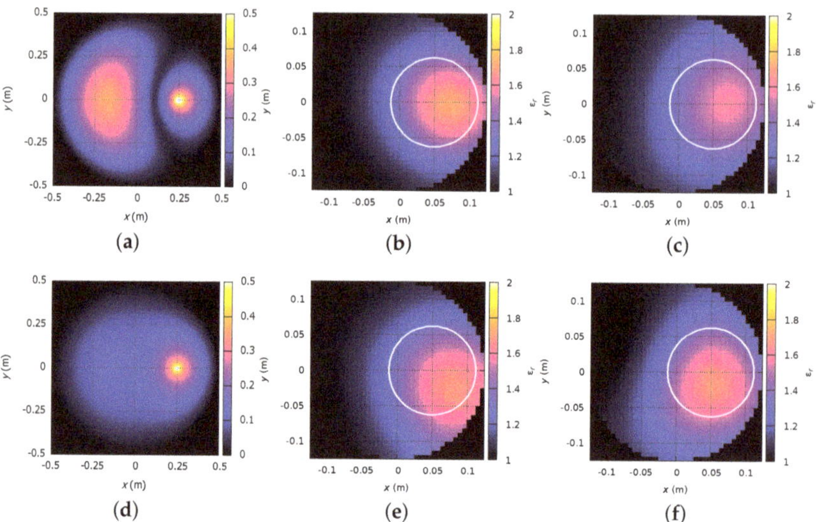

Figure 6. Examples of Green's functions and reconstructed distributions of the relative dielectric permittivity with small conducting enclosure, $a = \lambda/2$. Upper row: results with $\sigma_b = 1$ mS/m: (**a**) Green's function magnitude for a source at $(\rho_s, 0)$; (**b**) Reconstruction with cylindrical enclosure; (**c**) Reconstruction in free space. Lower row: results with $\sigma_b = 10$ mS/m: (**d**) Green's function for point source located at $(\rho_s, 0)$; (**e**) Reconstruction with cylindrical enclosure; (**f**) Reconstruction in free space.

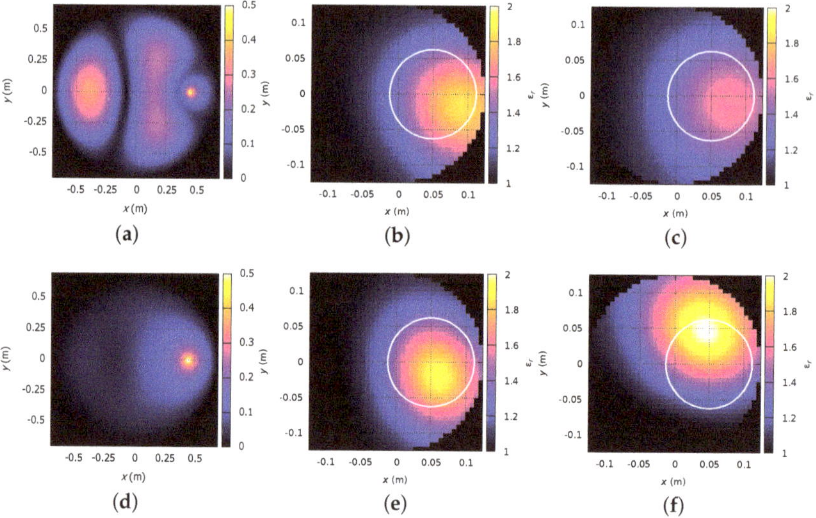

Figure 7. Examples of Green's functions and reconstructed distributions of the relative dielectric permittivity with small conducting enclosure, $a = 0.7\lambda$. Upper row: results with $\sigma_b = 1$ mS/m: (**a**) Green's function magnitude for a source at $(\rho_s, 0)$; (**b**) Reconstruction with cylindrical enclosure; (**c**) Reconstruction in free space. Lower row: results with $\sigma_b = 10$ mS/m: (**d**) Green's function for point source located at $(\rho_s, 0)$; (**e**) Reconstruction with cylindrical enclosure; (**f**) Reconstruction in free space.

As reported in the graphs, five different resonances may arise with this range of radiuses of the PEC cylinder. For small or absent background losses, the TM_{01}, TM_{11} and TM_{02} modes clearly cause a significant degradation of results, impairing a correct reconstruction of the inside dielectric cylinder. In the middle points between these significant resonances, results are comparable or even slightly better (in average) than the free space ones. When background losses rise, a completely different trend is observed (Figure 5c): there are no points with higher errors due to resonances, and the average of NRE demonstrates some advantages in adopting a conducting enclosure with respect to the free space conditions. Clearly, as expected, errors in both situations increase by enlarging the radius a, since this also increases the distance between the target and the observation domain O. However, the error increase with the surrounding metallic enclosure is slightly less steep than the free space case.

4.2. Large Conducting Enclosure

In the case of large conducting enclosure, the radius of the PEC cylinder has been varied in the range $a \in [1.0, 3.0]\lambda$ with 0.125λ steps. The same number of $S = 30$ positions as before has been used, but this time they are located on a circumference with radius $\rho_s = a - 0.25\lambda$. A multistatic and multiview configuration has been always considered. The investigation domain is larger than the previous case, and is a circular region centered at the origin with diameter $d_D = \lambda$. A circular dielectric cylinder characterized by the same dielectric properties as in Section 4.1, but with diameter $d_c = 0.2\lambda$ and centered at $(\rho_c, \theta_c) = (0.2\lambda, 0)$ is placed inside the PEC enclosure.

The forward electromagnetic problem has been solved by subdividing D into $N_f = 1976$ square subdomains with side length $l_{sf} = 0.02\lambda$, whereas a discretization with $N_i = 1264$ cells of dimension $l_{si} = 0.025\lambda$ has been used for the inverse problem's solution.

The mean value and standard deviation of the NRE in this case is shown in Figure 8, which reports the reconstruction errors versus the radius a for $\sigma_b = \{0, 1, 2\}$ mS/m $(\tan \delta = \{0, 0.06, 0.12\})$. Figure 9 shows some examples of the reconstructed relative dielectric permittivity inside D and the Green's function magnitude for a source located at $(\rho_s, 0)$. As can be noticed looking at Table 1, the number of resonances of the circular cavity dramatically increases for $a > \lambda$. This fact determines the significantly higher errors compared to the corresponding free space cases when no or low background loss is present. It is actually very difficult, even with a practical construction of a metallic chamber in mind, to avoid all the possible resonances. However, by increasing the background loss (e.g., Figure 8b) it can be observed that not all resonances are critical in the same way. Those that cause a worst degradation of reconstruction results are again related to the TM_{0l} modes, with $l = \{2, 3, 4, 5\}$, and can be seen in the high-error peaks at $a = \{1.375, 1.875, 2.375, 2.875\}\lambda$. Moreover, similarly to the case of the small enclosure, the TM_{1l} resonances have a smaller but yet significant impact. The remaining ones are sufficiently damped by the small background loss and do not produce increases in the NRE in neighboring sizes of the PEC cylinder. Of course, the bigger the cylinder radius, the greater is the effect of background losses.

Moreover, Figure 8c evidences some interesting facts related to the imaging in conducting enclosures versus their free space counterpart. In the lossy case, reconstruction results obtained inside conducting cylinders are always better then the free space ones, having a smaller mean NRE and even a significantly reduced standard deviation. This is also observed in Figures 9 and 10, where it can be seen that the free space reconstruction have more background artifacts than those obtained in conducting enclosures. In addition, the advantages of embedding the measurement configuration in a metallic cylinder appear more evident if the background losses rise.

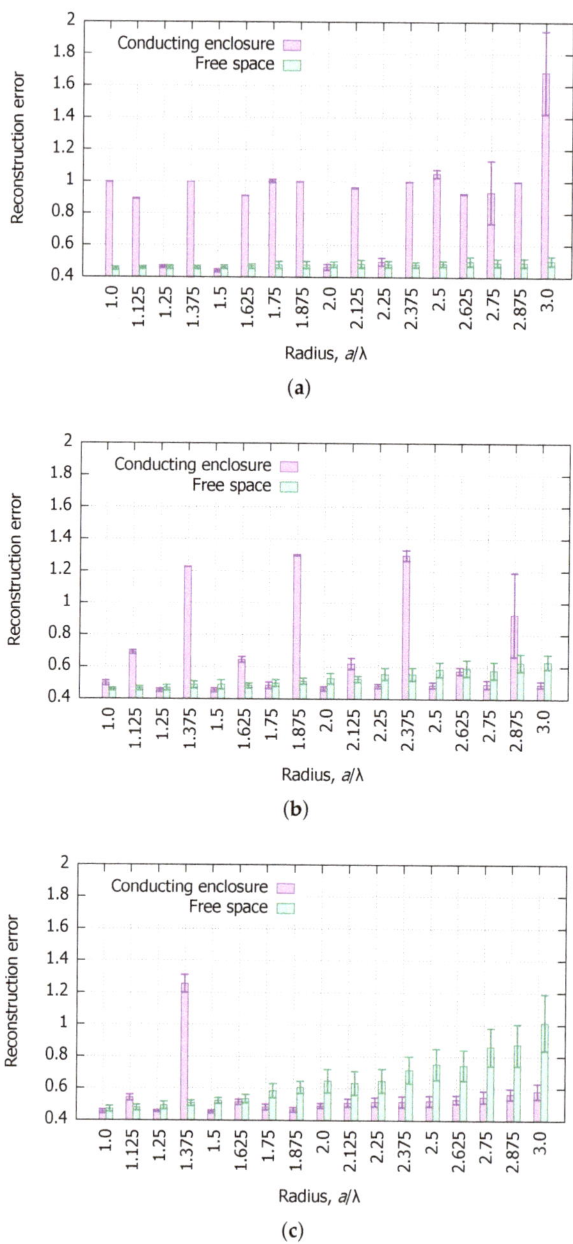

Figure 8. Large conducting enclosure. Reconstruction error versus the radius of the conducting enclosure: (**a**) $\sigma_b = 0$ S/m; (**b**) $\sigma_b = 1$ mS/m; (**c**) $\sigma_b = 2$ mS/m.

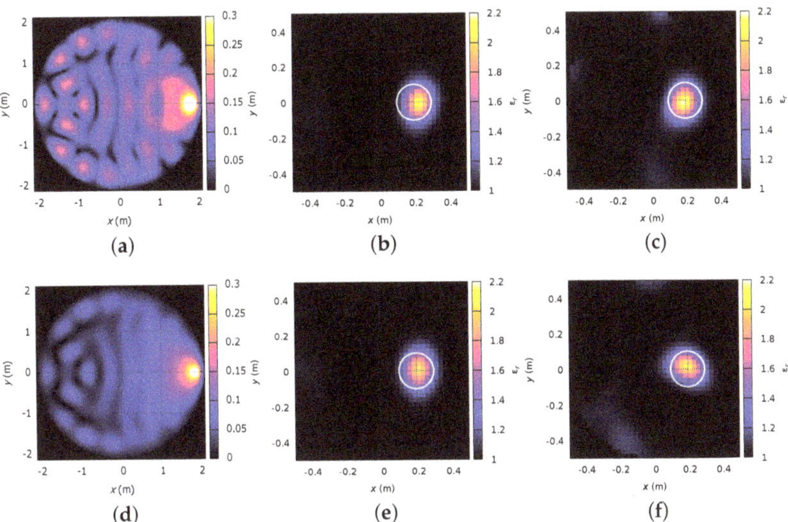

Figure 9. Examples of Green's functions and reconstructed distributions of the relative dielectric permittivity with small conducting enclosure, $a = 2.125\lambda$. Upper row: results with $\sigma_b = 1$ mS/m: (**a**) Green's function magnitude for a source at $(\rho_s, 0)$; (**b**) Reconstruction with cylindrical enclosure; (**c**) Reconstruction in free space. Lower row: results with $\sigma_b = 2$ mS/m: (**d**) Green's function for point source located at $(\rho_s, 0)$; (**e**) Reconstruction with cylindrical enclosure; (**f**) Reconstruction in free space.

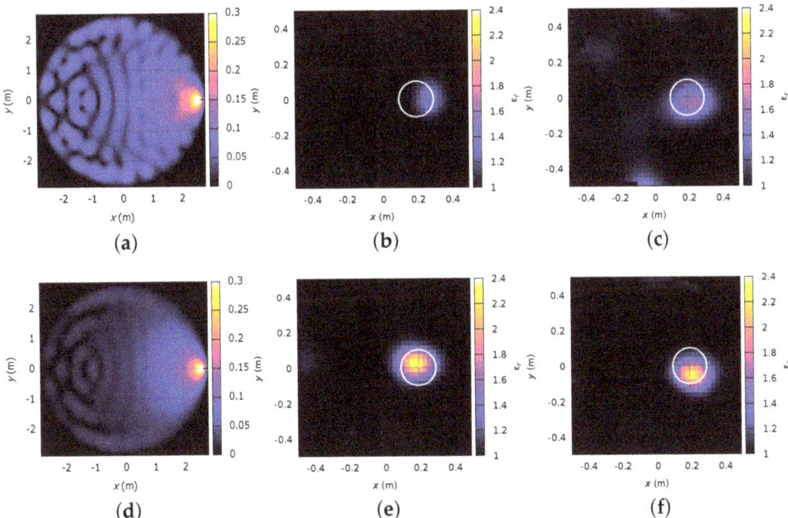

Figure 10. Examples of Green's functions and reconstructed distributions of the relative dielectric permittivity with small conducting enclosure, $a = 2.875\lambda$. Upper row: results with $\sigma_b = 1$ mS/m: (**a**) Green's function magnitude for a source at $(\rho_s, 0)$; (**b**) Reconstruction with cylindrical enclosure; (**c**) Reconstruction in free space. Lower row: results with $\sigma_b = 2$ mS/m: (**d**) Green's function for point source located at $(\rho_s, 0)$; (**e**) Reconstruction with cylindrical enclosure; (**f**) Reconstruction in free space.

In order to verify how such considerations generalize, additional tests with a different target have been carried out. In particular, a fixed imaging chamber size $a = 2.125\lambda$, with $\sigma_b = 2$ mS/m and all the other parameters as before has been considered. This time, the target is a cylinder with rectangular cross section of x, y side lengths equal to $d_{rx} = 0.15\lambda$ and $d_{ry} = 0.4\lambda$, respectively, centered at $(\rho_r, \theta_r) = (-0.2\sqrt{2}\lambda, 3\pi/4)$. It has a relative dielectric permittivity $\varepsilon_{r,r} = 2$ and electric conductivity $\sigma_r = 10$ mS/m. Some examples of the reconstructed dielectric profiles within the cylindrical enclosure and in free space have been shown in Figure 11a,b, respectively. Observing these images, it is confirmed that the PEC enclosure gives the best results even in this configuration, while the reconstruction obtained in free space is worse and more affected by background artifacts. Similar observations can be drawn from the reconstruction errors, reported in Table 2. As it happened with the circular target in the same conditions, both the mean value of the NRE and its standard deviation are significantly lower for the imaging inside cylindrical enclosure.

For comparison purposes, the same data have also been inverted by using an L^2 Hilbert-space formulation (i.e., with $p = 2$). Results are presented in Figure 11c,d. Clearly, in both PEC enclosure and free space, the dielectric properties of the target are underestimated. Furthermore, some artifacts appear in Figure 11d. The reconstruction errors listed in Table 2 lead to some interesting remarks. First, in all cases the proposed non-Hilbertian formulation gives by far the best results. Second, benefits of working inside the PEC enclosure still exist in the L^2 case, but are more evident when the regularization is performed outside Hilbert spaces. Third, the Hilbert-space NREs have a reduced standard deviation compared to the others, which is due to the higher smoothing effect. Although this may mitigate the impact of noise to some extent, it also degrades the reconstruction considerably (over-smoothing).

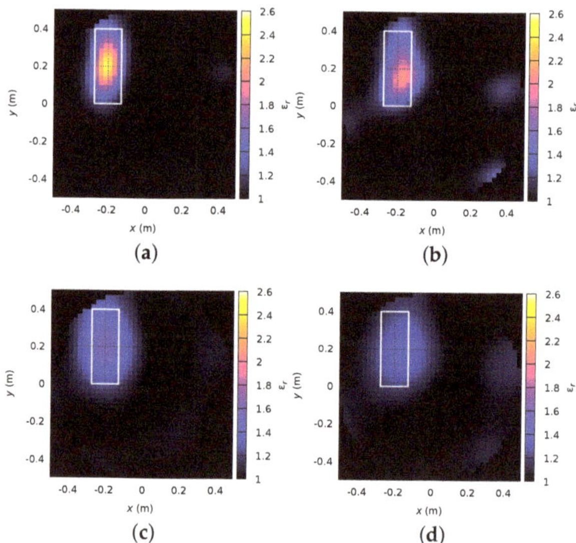

Figure 11. Rectangular cylinder. Examples of reconstructed distributions of the relative dielectric permittivity with enclosure of radius $a = 2.125\lambda$ and $\sigma_b = 2$ mS/m. Upper row: L^p-space results, with $p = 1.2$. (**a**) Reconstruction with cylindrical enclosure; (**b**) Reconstruction in free space. Lower row: Hilbert-space results. (**c**) Reconstruction with cylindrical enclosure; (**d**) Reconstruction in free space.

Table 2 also reports some computational data (time and random access memory) required to run the inversion methods on a personal computer equipped with an Intel(R) Core(TM) i5-2310 CPU at 2.90 GHz (Intel, Santa Clara, CA, USA) and 8 GB of RAM. Computational time is notably higher when a cylindrical enclosure is considered, and it shows a slight (average) increase in the proposed approach due to the number of performed Newton iterations. When the number of iterations is the same, there are no significant differences between methods. Furthermore, the present implementation of the case of PEC enclosure requires more random access memory than the free space case, but the amount of needed RAM is still similar for the L^p and the L^2 methods.

In summary, these results confirm the advantages of adopting a non-Hilbertian formulation for the imaging of targets inside conducting enclosures, and extend the previous analyses carried out in open space configurations [33].

Table 2. Rectangular cylinder. Reconstruction errors (NRE), computational times (mean value $\pm$ standard deviation) and required random access memory (RAM) with the proposed L^p-space approach compared with a Hilbert-space formulation.

		Proposed Approach	**Hilbert-Space**
	NRE	0.526 ± 0.020	0.706 ± 0.008
Cylindrical enclosure	Time (s)	90.24 ± 10.60	85.05 ± 0.203
	RAM (MB)	129.4 ± 0.254	129.5 ± 0.205
	NRE	0.622 ± 0.049	0.774 ± 0.020
Free space	Time (s)	12.34 ± 0.347	8.937 ± 2.695
	RAM (MB)	20.79 ± 0.233	20.78 ± 0.240

5. Conclusions

In this paper, the nonlinear microwave imaging of targets enclosed in circular conducting cylinders has been addressed. This particular imaging configuration, which presents some relevant potential applications in industrial nondestructive testing and medical diagnostics, has been considered within an inexact-Newton inversion method formulated in non-Hilbertian L^p spaces for the first time. The presence of the circular enclosure has been modeled with a proper Green's function as kernel of the integral operators used to formulate the imaging problem.

A brief discussion about the theoretical background has underlined that the problem faced in this paper is much more tricky than the analogous in free-space. Actually, while in free space the scattering field can be expanded into a simple sum of progressing waves, in the present problem the solution is made by a sum of complicate standing waves, and many resonant modes can arise inside the cavity. Furthermore, also the incident field is strongly affected by the cavity boundaries and contains many modal contributions, depending on the cavity dimensions.

The behavior of the proposed inverse scattering technique when operating in PEC enclosures has been analyzed from a numerical point of view in canonical configurations, involving circular dielectric cylinders. In particular, the simulations have been oriented at understanding the influence of the conducting enclosure on the reconstruction performance. Two distinct classes have been considered, regarding small and large enclosures, with different amounts of background loss.

As expected, for small enclosures without background losses, the first resonances cause a significant degradation of results, impairing a correct reconstruction of the inside dielectric cylinder. On the contrary, in the middle points between these significant resonances, results are comparable or even slightly better (in average) than the free space ones. Instead, when a slightly dissipative background is used, numerical experiments show a completely different behavior, since resonances are smoothed, and the average error seems to show some advantages in using a conducting enclosure with respect to the free space conditions. While, as expected, errors rise when enlarging the enclosure and moving measurement points away from targets, the error increase with the enclosure is slightly less steep than the free space case. Moreover, in the lossy case, reconstruction results obtained inside large conducting cylinders are always better then the free space ones, having a smaller mean error on the dielectric characterization of the targets and even a significantly reduced standard deviation. The advantages of embedding the measurement configuration in a metallic cylinder appear more evident if the background losses rise.

As it is well known, the selection of proper operating frequencies represents a crucial problem in time-harmonic microwave imaging. The presence of conducting enclosures further complicates this issue, due to the possible resonance phenomena. If the inversion method does not takes explicitly advantage from resonant modes, and low-loss configurations should be adopted, a general suggestion is to choose the working frequency so as to be sufficiently away from critical resonances, which can be computed based on the behavior of Bessel's functions. However, avoiding resonances may be very difficult as the size of the imaging chamber rises. As proven by the presented results, this problem is mitigated when the infill medium has non-negligible losses.

In conclusion, the adoption of a non-Hilbertian Lebesgue-space formulation seems promising for the imaging inside metallic cylinders, and may lead to more accurate dielectric reconstructions compared to the standard L^2 approach. Although some relevant differences may be observed based on the size of the outer metallic cylinder, ad-hoc chambers of appropriate dimensions can enhance results in the presence of lossy backgrounds. This could be exploited in many applications where not only boundaries can be controlled, but also the imaging devices can benefit from the use of circular enclosures, such as the case of biomedical imaging where the process requires a tight coupling of the investigated region with the surrounding background media.

Author Contributions: Conceptualization, A.F., M.P., A.R. and G.L.G.; methodology, A.F., M.P., A.R. and G.L.G.; formal analysis, A.F., M.P., A.R. and G.L.G.; investigation, A.F., M.P., A.R. and G.L.G. All authors have read and agreed to the published version of the manuscript.

Funding: This research received no external funding.

Data Availability Statement: The numerical data presented in this study are available from the corresponding author on request.

Conflicts of Interest: The authors declare no conflict of interest.

Abbreviations

The following abbreviations are used in this manuscript:

PEC	Perfect Electric Conductor
TM	Transverse Magnetic
NRE	Normalized Reconstruction Error
RAM	Random Access Memory

References

1. Kak, A.C.; Slaney, M. *Principles of Computerized Tomographic Imaging*; IEEE Press: New York, NY, USA, 1988.
2. Benedetto, A.; Pajewski, L. *Civil Engineering Applications of Ground Penetrating Radar*; Springer: Cham, Switzerland, 2015.
3. Bolomey, J.C. Advancing Microwave-Based Imaging Techniques for Medical Applications in the Wake of the 5G Revolution. In Proceedings of the 13th European Conference on Antennas and Propagation, Krakow, Poland, 31 March–5 April 2019; pp. 1–5.
4. Cakoni, F.; Colton, D. *Qualitative Methods in Inverse Scattering Theory: An Introduction*; Springer Science & Business Media: Berlin/Heidelberg, Germany, 2005.
5. Cakoni, F.; Colton, D.L.; Haddar, H. *Inverse Scattering Theory and Transmission Eigenvalues*; SIAM: Philadelphia, PA, USA, 2016.
6. Nikolova, N.K. *Introduction to Microwave Imaging*; EuMA High Frequency Technologies Series; Cambridge University Press: Cambridge, UK, 2017. [CrossRef]
7. Pastorino, M.; Randazzo, A. *Microwave Imaging Methods and Applications*; Artech House: Boston, MA, USA, 2018.
8. Crocco, L.; Litman, A. On embedded microwave imaging systems: Retrievable information and design guidelines. *Inverse Probl.* **2009**, *25*, 065001. [CrossRef]
9. Gilmore, C.; LoVetri, J. Enhancement of microwave tomography through the use of electrically conducting enclosures. *Inverse Probl.* **2008**, *24*, 035008. [CrossRef]

10. Coli, V.L.; Tournier, P.H.; Dolean, V.; Kanfoud, I.E.; Pichot, C.; Migliaccio, C.; Blanc-Féraud, L. Detection of simulated brain strokes using microwave tomography. *IEEE J. Electromagn. Microw. Med. Biol.* **2019**, *3*, 254–260. [CrossRef]

11. Gilmore, C.; Zakaria, A.; Pistorius, S.; LoVetri, J. Microwave Imaging of Human Forearms: Pilot Study and Image Enhancement. *Int. J. Biomed. Imaging* **2013**, *2013*, 673027. [CrossRef] [PubMed]

12. Asefi, M.; Baran, A.; LoVetri, J. An Experimental Phantom Study for Air-Based Quasi-Resonant Microwave Breast Imaging. *IEEE Trans. Microw. Theory Tech.* **2019**, *67*, 3946–3954. [CrossRef]

13. Fedeli, A.; Schenone, V.; Randazzo, A.; Pastorino, M.; Henriksson, T.; Semenov, S. Nonlinear S-parameters inversion for stroke imaging. *IEEE Trans. Microw. Theory Tech.* **2020**, in press. [CrossRef]

14. Winges, J.; Cerullo, L.; Rylander, T.; McKelvey, T.; Viberg, M. Compressed Sensing for the Detection and Positioning of Dielectric Objects Inside Metal Enclosures by Means of Microwave Measurements. *IEEE Trans. Microw. Theory Tech.* **2018**, *66*, 462–476. [CrossRef]

15. LoVetri, J.; Asefi, M.; Gilmore, C.; Jeffrey, I. Innovations in Electromagnetic Imaging Technology: The Stored-Grain-Monitoring Case. *IEEE Antennas Propag. Mag.* **2020**, *62*, 33–42. [CrossRef]

16. Chen, X.; Wei, Z.; Li, M.; Rocca, P. A review of deep learning approaches for inverse scattering problems. *Prog. Electromagn. Res.* **2020**, *167*, 67–81. [CrossRef]

17. Pavone, S.C.; Sorbello, G.; Di Donato, L. On the Orbital Angular Momentum Incident Fields in Linearized Microwave Imaging. *Sensors* **2020**, *20*, 1905. [CrossRef]

18. Bevacqua, M.T.; Isernia, T.; Palmeri, R.; Akinci, M.N.; Crocco, L. Physical insight unveils new imaging capabilities of orthogonality sampling method. *IEEE Trans. Antennas Propag.* **2020**, *68*, 4014–4021. [CrossRef]

19. Donelli, M.; Franceschini, D.; Rocca, P.; Massa, A. Three-dimensional microwave imaging problems solved through an efficient multiscaling particle swarm optimization. *IEEE Trans. Geosci. Remote Sens.* **2009**, *47*, 1467–1481. [CrossRef]

20. Fedeli, A.; Maffongelli, M.; Monleone, R.; Pagnamenta, C.; Pastorino, M.; Poretti, S.; Randazzo, A.; Salvadè, A. A tomograph prototype for quantitative microwave imaging: Preliminary experimental results. *J. Imaging* **2018**, *4*, 139. [CrossRef]

21. Donelli, M.; Manekiya, M.; Iannacci, J. Development of a MST sensor probe, based on a SP3T switch, for biomedical applications. *Microw. Opt. Technol. Lett.* **2021**, *63*, 82–90. [CrossRef]

22. Afsari, A.; Abbosh, A.M.; Rahmat-Samii, Y. Modified Born iterative method in medical electromagnetic tomography using magnetic field fluctuation contrast source operator. *IEEE Trans. Microw. Theory Tech.* **2019**, *67*, 454–463. [CrossRef]

23. Mojabi, P.; LoVetri, J. Eigenfunction contrast source inversion for circular metallic enclosures. *Inverse Probl.* **2010**, *26*, 025010. [CrossRef]

24. Abdollahi, N.; Jeffrey, I.; LoVetri, J. Non-Iterative Eigenfunction-Based Inversion (NIEI) Algorithm for 2D Helmholtz Equation. *Prog. Electromagn. Res. B* **2019**, *85*, 1–25. [CrossRef]

25. Rubek, T.; Meaney, P.M.; Meincke, P.; Paulsen, K.D. Nonlinear microwave imaging for breast-cancer screening using Gauss-Newton's method and the CGLS inversion algorithm. *IEEE Trans. Antennas Propag.* **2007**, *55*, 2320–2331. [CrossRef]

26. Mojabi, P.; LoVetri, J.; Shafai, L. A multiplicative regularized Gauss-Newton inversion for shape and location reconstruction. *IEEE Trans. Antennas Propag.* **2011**, *59*, 4790–4802. [CrossRef]

27. Abubakar, A.; Habashy, T.M.; Pan, G.; Li, M.K. Application of the multiplicative regularized Gauss-Newton algorithm for three-dimensional microwave imaging. *IEEE Trans. Antennas Propag.* **2012**, *60*, 2431–2441. [CrossRef]

28. Estatico, C.; Fedeli, A.; Pastorino, M.; Randazzo, A. Quantitative microwave imaging method in Lebesgue spaces with nonconstant exponents. *IEEE Trans. Antennas Propag.* **2018**, *66*, 7282–7294. [CrossRef]

29. Estatico, C.; Fedeli, A.; Pastorino, M.; Randazzo, A.; Tavanti, E. A phaseless microwave imaging approach based on a Lebesgue-space inversion algorithm. *IEEE Trans. Antennas Propag.* **2020**, *68*, 8091–8103. [CrossRef]

30. Schöpfer, F.; Louis, A.K.; Schuster, T. Nonlinear iterative methods for linear ill-posed problems in Banach spaces. *Inverse Probl.* **2006**, *22*, 311–329. [CrossRef]

31. Estatico, C.; Fedeli, A.; Pastorino, M.; Randazzo, A. Microwave imaging of elliptically shaped dielectric cylinders by means of an Lp Banach-space inversion algorithm. *Meas. Sci. Technol.* **2013**, *24*, 074017. [CrossRef]

32. Bisio, I.; Estatico, C.; Fedeli, A.; Lavagetto, F.; Pastorino, M.; Randazzo, A.; Sciarrone, A. Variable-exponent Lebesgue-space inversion for brain stroke microwave imaging. *IEEE Trans. Microw. Theory Tech.* **2020**, *68*, 1882–1895. [CrossRef]

33. Estatico, C.; Fedeli, A.; Pastorino, M.; Randazzo, A. Microwave imaging by means of Lebesgue-space inversion: An overview. *Electronics* **2019**, *8*, 945. [CrossRef]

34. Van Bladel, J.G. *Electromagnetic Fields*, 2nd ed.; IEEE Press Series on Electromagnetic Wave Theory; John Wiley & Sons: Hoboken, NJ, USA, 2007; Volume 19.

35. Balanis, C.A. *Advanced Engineering Electromagnetics*, 2nd ed.; John Wiley & Sons: Hoboken, NJ, USA, 2012.

36. Martinek, J.; Thielman, H.P. On Green's functions for the reduced wave equation in a circular annular domain with Dirichlet, Neumann and radiation type boundary conditions. *Appl. Sci. Res.* **1966**, *16*, 5–12. [CrossRef]

37. Duffy, D.G. *Green's Functions with Applications*, 1st ed.; Chapman & Hall/CRC: Boca Raton, FL, USA, 2001.

38. Kukla, S.; Siedlecka, U.; Zamorska, I. Green's functions for interior and exterior Helmholtz problems. *Sci. Res. Inst. Math. Comput. Sci.* **2012**, *11*, 53–62. [CrossRef]

39. Stokes, G.G. On the numerical Calculation of a Class of Definite Integrals and Infinite Series. *Trans. Camb. Philos. Soc.* **1856**, *9*, 166–187.

40. McMahon, J. On the Roots of the Bessel and Certain Related Functions. *Ann. Math.* **1894**, *9*, 23–30. [CrossRef]
41. Watson, G.N. The Zeros of Bessel Functions. *Proc. R. Soc. London. Ser. A Contain. Pap. Math. Phys. Character* **1918**, *94*, 190–206. [CrossRef]
42. Watson, G.N. *A Treatise on the Theory of Bessel Functions*, 2nd ed.; Cambridge University Press: Cambridge, UK, 1944; p. 804.
43. Elbert, Á.; Laforgia, A. An asymptotic relation for the zeros of Bessel functions. *J. Math. Anal. Appl.* **1984**, *98*, 502–511. [CrossRef]
44. Ifantis, E.; Siafarikas, P. Inequalities involving Bessel and modified Bessel functions. *J. Math. Anal. Appl.* **1990**, *147*, 214–227. [CrossRef]
45. Ifantis, E.; Siafarikas, P. Differential inequalities for the positive zeros of Bessel functions. *J. Comput. Appl. Math.* **1990**, *30*, 139–143. [CrossRef]
46. Ifantis, E.; Siafarikas, P. A differential inequality for the positive zeros of Bessel functions. *J. Comput. Appl. Math.* **1992**, *44*, 115–120. [CrossRef]
47. Breen, S. Uniform Upper and Lower Bounds on the Zeros of Bessel Functions of the First Kind. *J. Math. Anal. Appl.* **1995**, *196*, 1–17. [CrossRef]
48. Elbert, Á. Some recent results on the zeros of Bessel functions and orthogonal polynomials. *J. Comput. Appl. Math.* **2001**, *133*, 65–83. [CrossRef]
49. Segura, J. Bounds on Differences of Adjacent Zeros of Bessel Functions and Iterative Relations between Consecutive Zeros. *Math. Comput.* **2001**, *70*, 1205–1220. [CrossRef]
50. Pálmai, T.; Apagyi, B. Interlacing of positive real zeros of Bessel functions. *J. Math. Anal. Appl.* **2011**, *375*, 320–322. [CrossRef]
51. Kerimov, M.K. Studies on the zeros of Bessel functions and methods for their computation. *Comput. Math. Math. Phys.* **2014**, *54*, 1337–1388. [CrossRef]
52. Kokologiannaki, C.G.; Laforgia, A. Simple proofs of classical results on zeros of $J_\nu(x)$ and $J'_\nu(x)$. *Tbilisi Math. J.* **2014**, *7*, 35–39. [CrossRef]
53. Kerimov, M.K. Studies on the Zeroes of Bessel Functions and Methods for Their Computation: IV. Inequalities, Estimates, Expansions, etc., for Zeros of Bessel Functions. *Comput. Math. Math. Phys.* **2018**, *58*, 1–37. [CrossRef]
54. Joó, I. On the control of a circular membrane. I. *Acta Math. Hung.* **1993**, *61*, 303–325. [CrossRef]
55. Liu, H.; Zou, J. Zeros of the Bessel and spherical Bessel functions and their applications for uniqueness in inverse acoustic obstacle scattering. *IMA J. Appl. Math.* **2007**, *72*, 817–831. [CrossRef]
56. Kurup, D.G.; Koithyar, A. New Expansions of Bessel Functions of First Kind and Complex Argument. *IEEE Trans. Antennas Propag.* **2013**, *61*, 2708–2713. [CrossRef]
57. Beneventano, C.G.; Fialkovsky, I.V.; Santangelo, E.M. Zeros of combinations of Bessel functions and the mean charge of graphene nanodots. *Theor. Math. Phys.* **2016**, *187*, 497–510. [CrossRef]
58. Karamehmedović, M.; Kirkeby, A.; Knudsen, K. Stable source reconstruction from a finite number of measurements in the multi-frequency inverse source problem. *Inverse Probl.* **2018**, *34*, 065004. [CrossRef]
59. Qu, C.; Wong, R. "Best possible" upper and lower bounds for the zeros of the Bessel function $J_\nu(x)$. *Trans. Am. Math. Soc.* **1999**, *351*, 2833–2859. [CrossRef]
60. Ismail, M.E.; Muldoon, M.E. On the variation with respect to a parameter of zeros of Bessel and q-Bessel functions. *J. Math. Anal. Appl.* **1988**, *135*, 187–207. [CrossRef]
61. Yousif, H.A.; Melka, R. Bessel function of the first kind with complex argument. *Comput. Phys. Commun.* **1997**, *106*, 199–206. [CrossRef]
62. Doring, B. Complex Zeros of Cylinder Functions. *Math. Comput.* **1966**, *20*, 215–222. [CrossRef]

MDPI
St. Alban-Anlage 66
4052 Basel
Switzerland
Tel. +41 61 683 77 34
Fax +41 61 302 89 18
www.mdpi.com

Electronics Editorial Office
E-mail: electronics@mdpi.com
www.mdpi.com/journal/electronics

9 783036 522074